Melvin Calvin
Department of Chemistry
University of California
Berkeley, CA 94720

PHOTOPERCEPTION
BY PLANTS

PHOTOPERCEPTION BY PLANTS

PROCEEDINGS OF
A ROYAL SOCIETY DISCUSSION MEETING
HELD ON 9 AND 10 MARCH 1983

ORGANIZED AND EDITED BY
P. F. WAREING, F.R.S., AND H. SMITH

LONDON
THE ROYAL SOCIETY
1983

Printed in Great Britain for the Royal Society
at the
University Press, Cambridge

ISBN 0 85403 220 7

First published in *Philosophical Transactions of the Royal Society of London*,
series B, volume 303 (no. 1116), pages 345–536

CHEM

Published by the Royal Society
6 Carlton House Terrace, London SW1Y 5AG

PREFACE

The discovery of phytochrome by the classical researches of S. B. Hendricks and H. A. Borthwick and their associates in the 1950s constituted a landmark in the development of plant physiology, and it is now known that this substance plays a key role in all green plants. Phytochrome appears to be one important component of the machinery whereby green plants sense both quantitative and qualitative changes in the light conditions under which they grow. The other is a blue-light-absorbing photoreceptor, probably a flavoprotein – or family of flavoproteins – which is particularly involved in the responses of plants to directional variations in light intensity, i.e. phototropism. Together these two photosensors enable plants to perceive fluctuations in their radiation environment, including canopy shade and seasonal variations in daylength, i.e. photoperiodism. Phytochrome has been the more intensively studied, mainly because its unusual property of being able to undergo reversible photoconversion in response to variations in the spectral distribution of radiation to which the plant is exposed provides a specific and powerful experimental approach. However, the mechanisms whereby both of these photoreceptors transduce variations in the light environment to bring about growth and other responses remain unresolved and a matter of considerable debate.

During the 20 years since the existence of phytochrome was first demonstrated unequivocally, it has been the subject of intensive study and a large body of information on its biochemistry and physiology has been accumulated, and the acquisition of new knowledge continues at an unabated rate. More recently, data on the blue-absorbing photoreceptor have begun to accumulate. It is difficult for the non-specialist to keep abreast of progress in this rapidly moving field, so these proceedings of the Discussion Meeting on photoperception in plants, which brought together nearly all the world's leading authorities in this field and which was planned to provide a synthesis of current knowledge, will be particularly valuable to a wide range of professional plant physiologists and biochemists, in presenting an authoritative statement of 'the present state of the art'.

As organizers of the Discussion Meeting held in March 1983, we are grateful for the efficient assistance of the staff of the Royal Society, in particular Miss C. A. Johnson and Dr M. B. Goatly.

September 1983

H. Smith
P. F. Wareing

CONTENTS

CONTENTS

Phil. Trans. R. Soc. Lond. B **303**, 347–359 (1983)

Printed in Great Britain

Blue-light-absorbing photoreceptors in plants

By W. R. Briggs and M. Iino

*Department of Plant Biology, Carnegie Institution of Washington,
290 Panama Street, Stanford, California 94305, U.S.A.*

Evidence is presented that more than one blue-light photoreceptor plays a role in morphogenesis, and that there are at least three, distinguishable on the basis both of action spectra and other criteria, which may be found both in green plants and fungi. One of these has been tentatively identified as a flavoprotein–cytochrome complex, most probably located in the plasma membrane. Studies with oat seedlings suggest that it may be involved in photoreception for phototropism, at least for the first positive curvature response. Both photoreduction of the cytochrome, via excitation of the flavin, and phototropic sensitivity in the first positive curvature range are similarly affected by diphenylether herbicides. The second class of photoreceptors can be distinguished from the first in *Neurospora* by both genetic and physiological evidence, as well as by the action spectrum. It could be either flavin or carotenoid, although a different moiety is not excluded. The third class, distinguished only by action spectroscopy, shows a single sharp action peak near 475 nm, and seems unlikely to be either a flavin or a carotenoid, though they are not rigorously excluded. The first positive phototropic curvature response in maize shows a redistribution of growth consonant with the Cholodny–Went hypothesis for tropic responses, with an increase in the growth rate of the shaded side over dark controls, a concomitant decrease on the illuminated side, and no net change in overall growth rate.

Introduction

To understand any photosensory system, it is necessary to characterize the entire sequence of events from the primary photoact through to the final consequence of photoexcitation. The effects of blue light on higher plants and fungi are no exception. Light must first be absorbed by a photoreceptor moiety, leading to some primary photochemistry. The photochemistry, be it a photoreduction or oxidation, an isomerization, or some other type of reaction, then leads through a series of dark reactions to transduction of the light signal into a measurable physiological response. In only a very few systems (e.g. vision), do we have anything approaching complete information from identity of the photoreceptor, knowledge of its pertinent photochemistry, and knowledge of the transduction steps leading to the final response. Indeed in the blue light responses to be discussed below, we do not yet have definitive identification of the photoreceptor in any case, though in some cases the photoreceptor can be assigned to a particular class of pigments. Not surprisingly, in the absence of knowledge of the chemical nature of the photoreceptor it is difficult to unravel the transduction steps leading to the response.

In blue-light-sensitive systems, efforts have been expended either at the photoreceptor end of the chain (action spectroscopy, inhibitor studies, pigment isolation attempts, etc.) or the response end (e.g. the relation between auxin distribution and tropic responses), with few studies in between. This paper will deal primarily with the question of photoreceptor

27-2

characterization in several systems, emphasizing three systems that may be shared between green plants and fungi. It will then consider briefly the response mechanism in one particular case. It is not our purpose here to be comprehensive. For more detailed coverage, the reader is referred to the proceedings of a recent symposium (Senger 1980) and several recent reviews. Senger (1982) and Senger & Briggs (1981) provide general treatments, Tan (1978) treats the filamentous fungi in detail, and Lenci & Columbetti (1978) deal with the photoresponses of a broad range of microorganisms.

RESPONSES IN WHICH A FLAVOPROTEIN IS IMPLICATED AS PHOTORECEPTOR

Action spectroscopy

A number of blue-light photoresponses have been described that have an action spectrum highly suggestive of the absorption spectrum of a flavoprotein. These action spectra all show maximum activity between 400 and 500 nm, with fine structure revealing at least three bands, the largest of which is between 450 and 460 nm. They also show a single somewhat smaller

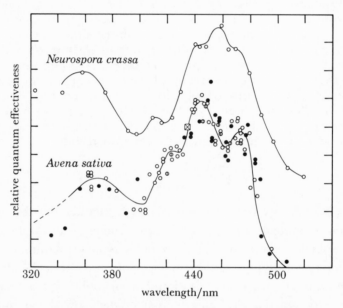

FIGURE 1. Action spectra for first positive phototropic curvature of *Avena* coleoptiles (lower curve, redrawn after Thimann & Curry (1960)), and for photoreduction *in vivo* of a *b*-type cytochrome in *Neurospora* mycelium (upper curve, redrawn after Muñoz & Butler (1965)).

broad band of activity in the near ultraviolet, centred near 360 nm. Among such photoresponses are phototropic curvature of etiolated *Avena sativa* coleoptiles (Shropshire & Withrow 1958; Thimann & Curry 1960) (figure 1, lower curve); phototropic curvature of *Phycomyces* sporangiophores (Curry & Gruen 1959); light-induced carotenogenesis in *Fusarium aquaeductuum* (Rau 1967); and suppression of a circadian rhythm of conidiation in a mutant strain of *Neurospora crassa* (Sargent & Briggs 1966). There are a number of other such responses, several of them treated in detail by Tan (1978). All of these action spectra resemble fairly closely the absorption spectrum of a typical flavoprotein, e.g. NADPH–cytochrome *c* oxidoreductase (Kamin *et al.* 1966), leading to the hypothesis that the flavin moiety of a flavoprotein is the photoreceptor. Because the absorption spectra of most flavoproteins are extremely similar, however, the action spectra cited above are of little assistance in indicating which flavoprotein might be the best candidate.

Light-induced flavin-mediated cytochrome reduction

About 10 years ago, Butler and his colleagues first detected blue-light-induced absorbance changes *in vivo* in several fungi (see Senger & Briggs (1981) for review). The careful studies by Muñoz & Butler (1975) showed that the light-induced absorbance change in *Neurospora crassa* mycelium involved the photoreduction of a *b*-type cytochrome, with the simultaneous reduction of a flavoprotein. Both constituents became reoxidized in the dark. The action spectrum for the process (figure 1, upper curve) was very much like that for the various processes described in the previous section. Brain *et al.* (1977a) subsequently found the photoactive cytochrome–flavin to be located in the plasma membrane of *Neurospora* and showed similar photoactivity in a membrane fraction from corn coleoptiles. Goldsmith *et al.* (1980) improved the techniques for measuring the photoreaction so that it could be induced consistently in corn membrane fractions, and showed that the flavin moiety was firmly bound. They also presented spectral evidence that the photoreaction was specific only for a *b*-type cytochrome with a room temperature α band near 560 nm. This specificity was obtained, despite the presence of both cytochrome b_5 from the endoplasmic reticulum and significant mitochondrial contamination. The specificity was also obtained whether the photosensitizer was the endogenous flavin moiety or externally added flavins or methylene blue. Exogenous photosensitizers could be readily washed away, but the endogenous photoreceptor for the photoreduction was firmly membrane-bound. Leong & Briggs (1981) used sucrose and renografin gradient centrifugation to separate the membrane fraction containing the photoactivity from mitochondria, endoplasmic reticulum, and Golgi, and showed that it cosedimented with glucan synthetase II activity. It seems likely that as with *Neurospora* (Brain *et al.* 1977a), this fraction is also the plasma membrane (see Quail (1979) for an evaluation of the evidence that glucan synthetase II is a reliable marker for plasma membrane in higher plants). Leong & Briggs (1981) also reported that the complex could be separated from the bulk of the membrane fraction by mild detergent treatment with little loss of photoactivity. Hence it was probably associated with a peripheral membrane protein rather than an intrinsic one. Finally, Leong *et al.* (1981) obtained reduced-minus-oxidized spectra for the cytochrome at liquid nitrogen temperature. The single sharp α band at 555 nm clearly distinguishes this *b*-type cytochrome both from the cytochrome b_5 of the endoplasmic reticulum, with a split α band showing peaks at 550 and 556 nm, and the several mitochondrial cytochromes. Leong *et al.* (1981) also determined that the redox potential of the cytochrome at 50 % reduction was -65 mV.

There are several lines of evidence that the photoreceptor in the complex is actually a flavin. First, as mentioned above, the action spectrum of the cytochrome photoreduction, at least in *Neurospora*, resembles the absorption spectrum of a flavoprotein (Muñoz & Butler 1975). Second, the purified photoactive membrane fraction clearly contains a flavin: the fluorescence excitation spectrum for emission at 525 nm is clearly that of a flavoprotein (Leong *et al.* 1981). Third, phenylacetic acid, which reacts with flavins in their triplet state to form a covalent linkage with them (Hemmerich *et al.* 1967), and iodide and azide, both effective in depopulating the excited states of flavins (Heelis *et al.* 1978), are effective inhibitors of the blue-light-induced photoreduction (Caubergs *et al.* 1979).

[3]

Relevance of flavin-mediated cytochrome photoreduction to photoresponses in vivo

None of the evidence so far cited provides any convincing argument that this photosensitive flavoprotein–cytochrome complex is indeed the biologically active photoreceptor for any of the responses cited. There is, however, accumulating evidence, both for coleoptile phototropism and the conidiation response in *Neurospora*, that this complex may indeed be the actual photoreceptor. First, the same three inhibitors that were effective in blocking the photoreduction *in vivo* (phenylacetic acid, iodide and azide (see Caubergs *et al.* 1979)) preferentially inhibited the phototropic over the geotropic response of corn coleoptiles (Schmidt *et al.* 1977). Second, a mutant of *Neurospora*, *poky*, which is deficient in all of its *b*-type cytochrome, has only about 1 % of the wild-type photoreducible cytochrome activity, and is less than 1 % as sensitive as wild-type to photosuppression of the circadian conidiation pattern (Brain *et al.* 1977 *b*). Third, methylene blue, which photosensitizes reduction of the *b*-type cytochrome to red light (Britz *et al.* 1979), also sensitizes photosuppression of the conidiation banding in *Neurospora* and phase-shifting of the banding (probably simply a special case of suppression) in response to red light (see Briggs 1980). Finally, Leong & Briggs (1982) showed that the diphenyl ether herbicide acifluorfen both increased the level of photoreduction of the *b*-type cytochrome in membrane fractions from *Avena* coleoptiles and significantly sensitized the first positive phototropic curvature reaction. The herbicide was entirely without effect on the geotropic reaction. While not one of these lines of evidence is more than circumstantial, and each case requires further careful analysis, they, together with the action spectra mentioned above, make the hypothesis that this light-sensitive plasma-membrane-associated flavin–cytochrome complex is the actual photoreceptor for phototropism in coleoptiles and for photosuppression of conidial banding in *Neurospora* an attractive one.

RESPONSES IN WHICH EITHER A FLAVIN OR CAROTENOID MIGHT BE THE PHOTORECEPTOR

Carotenogenesis in Neurospora crassa

Many years ago, Zalokar (1955) first published an action spectrum between 400 and 500 nm for carotenogenesis in *Neurospora*. Although frequently included in collections of action spectra thought to share a common photoreceptor (Muñoz & Butler 1975), in particular those discussed above that are hypothesized to have a flavoprotein as photoreceptor, there are features of this action spectrum that are not consistent with those mentioned above. First, the wavelength for maximum photoactivity is much nearer 480 nm than 450–460 nm; and second, the relative activity at 400 nm with respect to the maximum is considerably lower than in the spectra mentioned above. A more recent detailed action spectrum for this process (De Fabo *et al.* 1976) confirms and extends these conclusions (figure 2, upper curve). There are two sharp peaks, one at 450 nm and the other at 480 nm, with the latter being the highest. Relative activity is indeed low, though not absent, between 300 and 400 nm.

There are currently three lines of evidence suggesting that the photoreceptor for carotenogenesis in *Neurospora* is not identical with that for the suppression of conidial banding. The first of these involves study of the *poky* mutant. E. Schrott (unpublished) found that although, as mentioned above, the deficiency in *b*-type cytochromes brought about a reduction to less than one-hundredth in light sensitivity for the suppression of conidial banding, it was entirely without effect on carotenogenesis: the dose–response curve for carotenogenesis in the *poky* mutation was indistinguishable from that of the wild-type. The second line of evidence involved studies with an external photosensitizer. Schrott (unpublished) showed that although methylene blue would sensitize both the suppression of conidial banding and phase-shifting of the circadian rhythm of conidiation to red light (see above), red light administered in the presence of methylene blue failed to have any effect whatever on carotenogenesis either in *poky* or the wild-type. The third line of evidence comes from studies by Paietta & Sargent (1981) with two riboflavin mutants,

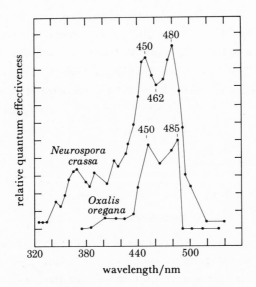

FIGURE 2. Action spectra for carotenogenesis in *Neurospora* (upper curve, redrawn after De Fabo *et al.* (1976)), and light-induced leaflet closure in *Oxalis oregana* (lower curve, redrawn after Björkman & Powles (1981)).

rib-1 and *rib-2*, of *Neurospora*. Flavin deficiency reduced light sensitivity for suppression of conidial banding to just over one-hundredth, whereas it reduced light sensitivity for carotenogenesis to only about one-quarter.

These results, taken together, strongly suggest that there are two different blue-light photoreceptors in *Neurospora*, one probably mediated by a light-sensitive flavin–cytochrome complex localized in the plasma membrane, and the other clearly not. The second photoreceptor could be a flavin, a carotenoid, or something else. There is at present no evidence other than action spectroscopy bearing on its chemical nature, and that evidence is ambiguous.

Blue-light-induced leaflet closure in Oxalis oregana

Björkman & Powles (1981) have recently described a rapid light-induced closure of the leaflets of *Oxalis oregana*, a ground cover species commonly found deep in California's redwood forests. These plants normally flourish in deep shade, and are very well adapted to these shade conditions. Exposure to light fluence rates equivalent to full sunlight leads within minutes to significant damage to the photosynthetic apparatus, and prolonged exposures lead to damage that requires hours of recovery, or that indeed may be permanent. Even in the deepest shade of the forest floor, however, there are occasional sunflecks in which the light fluence rate approaches or equals that of unobstructed sunlight. The fluence rate may increase as much as 200-fold within 1 min or less. In Nature, the plants respond by rapid folding of the leaflets so that only their edges are exposed to the full sunligt. With a very much reduced area of light interception for the same amount of chlorophyll, photoinhibition is completely avoided. The reaction begins within seconds of the arrival of a sunfleck as the Earth rotates, and goes almost to completion within as little as 3 min. On departure of the sunfleck, the leaflets then require approximately $\frac{1}{2}$ h to return to their normal extended orientation.

The action spectrum for this response (Björkman & Powles 1981) is shown in figure 2 (lower curve), along with that for carotenogenesis in *Neurospora* (figure 2, upper curve). Like the *Neurospora* spectrum, that for the *Oxalis* response shows two distinct peaks in the visible, one near 450 nm and the other near 480 nm (at a slightly longer wavelength for *Oxalis* than for *Neurospora*), and low activity in the near ultraviolet (lower in *Oxalis* than in *Neurospora*). Finally, both spectra share a minimum at about 460 nm. Though a flavin is not excluded in either case, the low activity in the near ultraviolet makes a flavin a somewhat unlikely candidate, and a carotenoid more likely. Even if the photoreceptors were flavins, it is highly unlikely that the flavins would be the same as those involved in the other responses so far discussed.

RESPONSES WHERE NEITHER A FLAVIN NOR A CAROTENOID IS A PROMISING CANDIDATE FOR THE PHOTORECEPTOR

Responses in fungi

In addition to the blue light responses discussed above, there are several that have action spectra showing only a single sharp action peak near 475 nm, and no other significant fine structure. Among these is that for the inhibition of conidiation in the fungus *Stemphylium botryosum* (Leach 1968), and that for promotion of sporulation in *Penicillium isariiforme* (Benninck 1971). That for *Stemphylium* is shown in figure 3 (lower curve). In comparison with the two

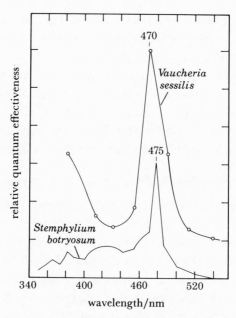

FIGURE 3. Action spectra for inhibition of conidiation in the fungus *Stemphylium botryosum* (lower curve, redrawn after Leach (1968)), and for light-induced chloroplast aggregation in the alga *Vaucheria sessilis* (upper curve, redrawn after Blatt (1980)).

responses discussed in the preceding section, a flavin as photoreceptor in these cases seems even more unlikely. It is also hard to invoke a carotenoid when the action peak is as narrow as these are.

Chloroplast aggregation in Vaucheria sessilis

The final process to be discussed is the light-induced aggregation of chloroplasts in the alga *Vaucheria sessilis*. Blatt (1980) measured the rates of chloroplast aggregation as a function of fluence rate for a series of wavelengths between 385 and 528 nm. He had previously shown that there was no effect of red light on this system. The action spectrum for the relative quantum efficiency as a function of wavelength is shown in figure 3 (upper curve). As with the two fungal responses just mentioned, the action spectrum shows only a single peak near 475 nm. *Vaucheria* does show somewhat more sensitivity in the near ultraviolet than either of the fungal responses. The requirement for using fibre optics both for excitation and measuring beams precluded measurement of the response below 385 nm, because adequate quantum fluences could not be obtained.

ON THE MULTIPLICITY OF BLUE-LIGHT PHOTORECEPTORS

It should be clear from the above discussion that there must be more than one blue-light photoreceptor both in fungi and in green plants. This idea is hardly original, as Tan (1978) stresses it for the filamentous fungi, and both Shropshire (1980) and De Fabo (1980) and others have addressed this possibility. Even the notion that there may be two different blue-light photoreceptors in the same organism, as may occur in *Neurospora*, is not new. Mikolajaczyk & Diehn (1975) proposed that more than one pigment is involved in the photophobic responses of *Euglena* to blue light. One of these photoreceptors showed strong inhibition in the presence of potassium iodide, and hence could have been a flavin; the other was iodide-insensitive. In addition, Jayaram *et al.* (1979) present arguments favouring two different photoreceptors for carotenogenesis in *Phycomyces*. In this connection, it is of some interest that the action spectrum for photoinduction of carotenoid synthesis for the water mould *Fusarium aquaeductuum* (Rau 1967) is similar to that for flavin-mediated photoreduction of the membrane-associated *b*-type cytochrome in *Neurospora* (Muñoz & Butler 1975); and both the *Fusarium* response (Lang-Feulner & Rau 1975) and the cytochrome photoreduction (Britz *et al.* 1979) can be potentiated by red light if methylene blue is present. By contrast, the action spectrum for carotenogenesis in *Neurospora* itself (De Fabo *et al.* 1976) does *not* resemble that for photoreduction of the *Neurospora* cytochrome, and the process is *not* potentiated by red light in the presence of methylene blue. Hence not only may *Neurospora* contain two different photoreceptors, but a single process in the filamentous fungi, namely carotenogenesis, may be potentiated by different photoreceptor molecules in different organisms.

In summary, a review of a number of different action spectra suggests that the fungi may share at least three classes of photoreceptors with their photosynthetic cousins. One of these classes is clearly a flavin, and at least in the phototropism of corn and oat coleoptiles and the photosuppression of conidial banding in *Neurospora* it may involve a plasma membrane-associated flavin–cytochrome complex as the photoreceptor. A second class, that associated with carotenogenesis in *Neurospora* and leaflet closure in *Oxalis*, could be a flavin, a carotenoid, or some other compound. If this class is a flavin, it is unlikely to be the same one as involved in the first class, because, at least in *Neurospora*, a *b*-type cytochrome is evidently not involved. The third class, found in some filamentous fungi and, at least in *Vaucheria*, among green plants, shows just a single sharp action peak in the blue. It seems unlikely that it could be either a flavin or a carotenoid, though neither can be completely ruled out.

Action spectra do not, of course, necessarily reflect the absorption spectrum of the photoreceptor pigment. They may be altered by screening pigments or may be a consequence of photochromicity, reflecting the absorption spectrum of neither member of the photochromic pair of molecular species directly (see Shropshire 1980). Because the two photosystems in *Neurospora* can be distinguished on evidence other than action spectroscopy, it is immaterial to the hypothesis that there are two different photoreceptors whether the action spectra precisely resemble the photoreceptor absorption spectra or not. Furthermore, as Shrophire (1980) points out, the participation of a screening pigment in the carotenogenesis response is unlikely because the irradiations were done through a single mycelial pad, and what little absorbance was found was insufficient to modify the action spectrum in any significant way. A role of screening is likewise unlikely for suppression of conidial banding (Sargent & Briggs 1966) for similar reasons. Screening is also most unlikely to be the cause of the single sharp action peak near 475 nm in *Vaucheria*: the photoreceptor is located in the plasma membrane or outermost cortical cytoplasm (Fischer-Arnold 1963), with only the cell wall intervening between incident light and photoreceptor. Photochromicity cannot at present be ruled out, though if cycling of a photochromic pigment is required for action, a mechanism that *could* produce the sharp peak, this system must differ from the others described, where a mechanism requiring such cycling for action cannot be readily invoked

THE TRANSDUCTION STEP IN COLEOPTILE PHOTOTROPISM

The Went–Cholodny hypothesis for tropisms states that the tropisms are the consequence of a light-induced or gravity-induced lateral redistribution of auxin, leading to a redistribution in growth, causing the directed growth response (see Went & Thimann 1937). All of the early studies were based on experiments with grass coleoptiles, and later studies adding support to the hypothesis for gravitropism (Gillespie & Briggs 1961; Gillespie & Thimann 1963) and phototropism (Briggs *et al.* 1957; Briggs 1963; Pickard & Thimann 1964) also involved auxin studies with coleoptiles. These more recent studies clearly demonstrated that gravitational or light stimuli leading to tropic curvature in coleoptiles were indeed accompanied by lateral transport of auxin, and that neither photoinactivation of auxin nor inhibition of auxin synthesis by light could account for the auxin differential.

Firn & Digby (1980) have raised some provocative questions concerning the Went–Cholodny hypothesis. They point out first that if the hypothesis is really valid, one would expect an increase in the growth rate of the shaded (positive phototropism) or lower (gravitropism) side of the coleoptile, with respect to an unstimulated control, and a corresponding decrease in growth on the opposite side. They also point out that for the auxin differential to be responsible for the growth curvature, its formation would have to *precede* the actual changes in growth rate. Recently Franssen *et al.* (1981, 1982) have reported that continuous blue light administered under conditions of fluence rate and duration sufficient to yield second positive curvature (see Zimmerman & Briggs 1963) caused a complete cessation of growth on the illuminated side of *Avena* coleoptiles, with little change in the growth rate of the shaded side, a result clearly not consistent with the Went–Cholodny hypothesis.

There is accumulating evidence that there are phytochrome responses in corn and oat seedlings that are so sensitive that even short exposures (minutes or even seconds) of dim green or blue light can cause substantial growth changes (Blaauw *et al.* 1968; Briggs & Chon 1966; Chon & Briggs 1966; Iino 1982*a*, *b*; Iino & Carr 1981; Mandoli & Briggs 1981). It therefore seems possible that either the blue light used for phototropic induction or the green safelight used for handling the plants could bring about growth responses that were unrelated to the tropic responses, but which could easily obscure the growth differential which was the actual basis for the tropic response. Indeed, Pratt & Briggs (1966) showed spectrophotometrically that both blue and green light could bring about significant P_r–P_{fr} phototransformation *in vivo* in corn [maize] coleoptiles.

We therefore set out to measure growth changes brought about by blue light dosages leading to first positive curvature under conditions under which all phytochrome reactions would be

TABLE 1. GROWTH REDISTRIBUTION DURING FIRST POSITIVE CURVATURE OF CORN COLEOPTILE

(Curvature was developed for 100 min after the onset of illumination.)

$\log_{10}\{\text{fluence}/(\mu\text{mol m}^{-2})\}$	curvature/deg	growth change (% of control) illuminated side	shaded side
−1.0	2	−5	5
−0.2	12	−30	30
0.7	24	−45	55
1.6	12	−15	40
2.2	8	0	35

saturated. In brief, corn seedlings were raised under continuous red light ($0.15 \mu\text{mol m}^{-2}\text{ s}^{-1}$) at 24–25 °C for 3 days from the time of sowing. Phototropic induction with blue light and subsequent curvature development also took place under red light. In this way, we hoped to minimize any effects of phytochrome phototransformation by the blue light itself. The reasoning was that the very low fluences of blue light used for phototropic induction would have a negligible effect on the phytochrome photoequilibrium, which was sustained by the high fluence-rate red light. The phototropic fluence–response curve so obtained yielded the same threshold and saturation values as those obtained in 1966 by Chon & Briggs with dark-grown corn seedlings given only a brief exposure to red light 2 h before phototropic induction.

For growth measurements, each coleoptile was carefully marked just before phototropic induction exactly 1.5 cm below the coleoptile apex. An image of a set of plants was obtained by photocopying, and the photocopies used to measure the starting lengths of the sides that would be illuminated or shaded in the experimental plants. After curvature had developed for 100 min after the onset of blue-light treatment, the experimental plants were also photocopied for subsequent analysis. Length measurements were obtained with a digitizer directly on line with a computer. Growth increments were obtained for both shaded and illuminated sides, and the results expressed as percentage of those of control plants, which had received no phototropic induction (table 1).

The results obtained over the entire first positive curvature range showed an increase in growth on the shaded side of the coleoptile, and a compensatory decrease on the illuminated side: precisely what one would expect if the Went–Cholodny hypothesis were valid. The relative size of the increase or decrease is directly related to the amount of curvature obtained, with maximal curvature being accompanied by the maximal differential. At the higher fluences, there was some evidence for an overall stimulation of growth, but this tendency was eliminated when just the coleoptile tips were irradiated with blue light (data not shown). The kinetics of the differential growth response were also obtained. For these measurements, enlargements of photographic images of the coleoptiles, obtained with red-sensitive film, were used to increase resolution. The results obtained for maximal first positive curvature showed a clear stimulation of the growth rate of the shaded side and a decrease in the growth rate on the illuminated side. Net growth was hardly changed. These results will be presented in detail elsewhere.

We hope to begin studies on the kinetics of auxin redistribution soon, in order to address the second point raised by Firn & Digby (1980). We also hope to obtain measurements in the second positive curvature range to determine whether the mechanism in that range is really different, as suggested by the experiments of Frannsen et al. (1981, 1982), or whether the difference found can be attributed to light reactions unrelated to the direct physiological consequence of photostimulation of the blue-light photoreceptor.

Concluding remarks

It is clear that there are complexities involved at both the photoreceptor and response ends of the blue-light-excited sensory transduction chain. At the photoreceptor end, hypotheses based on the expectation of one kind of photoreceptor may be invalid because quite another photoreceptor is involved. At the response end, the interaction of photoreceptors may obscure a response mechanism, or different types of responses (e.g. first as against second positive curvature in coleoptiles or coleoptile tropisms compared with those of hypocotyls) may involve different mechanisms. Further experimentation is needed to resolve these problems, and is certainly required before any intermediate steps in photosensory transduction in response to blue light can be characterized in any of these systems.

This is Carnegie Institution of Washington Department of Plant Biology Publication no. 816.

References

Benninck, G. J. H. 1971 Photomorphogenese bij *Penicillium isariiforme*. Ph.D. thesis, University of Amsterdam.

Björkman, O. & Powles, S. B. 1981 Leaf movement in the shade species *Oxalis oregana*. I. Response to light level and light quality. *Carnegie Instn Wash. Yb.* **80**, 59–62.

Blaauw, O. H., Blaauw-Jansen, G. & van Leeuwen, W. J. 1968 An irreversible red-light-induced growth response in *Avena*. *Planta* **82**, 87–104.

Blatt, M. R. 1980 A study of events associated with light-dependent chloroplast movement in the alga *Vaucheria sessilis*. Ph.D. thesis, Stanford University, California.

Brain, R. D., Freeberg, J., Weiss, C. V. & Briggs, W. R. 1977a Blue light-induced absorbance changes in membrane fractions from corn and *Neurospora*. *Pl. Physiol.* **59**, 948–952.

Brain, R. D., Woodward, D. O. & Briggs, W. R. 1977b Correlative studies of light sensitivity and cytochrome content in *Neurospora crassa*. *Carnegie Instn Wash. Yb.* **76**, 295–299.

Briggs, W. R. 1963 Mediation of phototropic responses of corn coleoptiles by lateral transport of auxin. *Pl. Physiol.* **38**, 237–247.

Briggs, W. R. 1980 A blue light photoreceptor system in higher plants and fungi. In *Photoreceptors and plant development* (ed. J. De Greef), pp. 17–28. Antwerpen: University of Antwerpen Press.

Briggs, W. R. & Chon, H. P. 1966 The physiological versus the spectrophotometric status of phytochrome in corn coleoptiles. *Pl. Physiol.* **41**, 1159–1166.

Briggs, W. R., Tocher, R. D. & Wilson, J. F. 1957 Phototropic auxin redistribution in corn coleoptiles. *Science, Wash.* **126**, 210–212.

Britz, S. J., Schrott, E., Widell, S. & Briggs, W. R. 1979 Red light-induced reduction of a particle-associated *b*-type cytochrome from corn in the presence of methylene blue. *Photochem. Photobiol.* **29**, 359–366.

Caubergs, R. J., Goldsmith, M. H. M. & Briggs, W. R. 1979 Light-inducible cytochrome reduction in membranes from corn coleoptiles: fractionation and inhibitor studies. *Carnegie Instn Wash. Yb.* **78**, 121–125.

Chon, H. P. & Briggs, W. R. 1966 The effect of red light on the phototropic sensitivity of corn coleoptiles. *Pl. Physiol.* **41**, 1715–1724.

Curry, G. M. & Gruen, H. E. 1959 Action spectra for the positive and negative phototropism of *Phycomyces* sporangiophores. *Proc. natn. Acad. Sci. U.S.A.* **45**, 797–804.

De Fabo, E. C. 1980 On the nature of the blue light photoreceptor: still an open question. In *The blue light syndrome* (ed. H. Senger), pp. 185–197. Berlin, Heidelberg and New York: Springer-Verlag.

De Fabo, E. C., Harding, R. W. & Shropshire, W. Jr 1976 Action spectrum between 260 and 800 nanometers for the photoinduction of carotenoid biosynthesis in *Neurospora crassa*. *Pl. Physiol.* **57**, 440–445.

Firn, R. D. & Digby, J. 1980 The establishment of tropic curvatures in plants. *A. Rev. Pl. Physiol.* **31**, 131–148.

Fischer-Arnold, G. 1963 Untersuchungen über die Chloroplastenbewegung bei *Vaucheria sessilis*. *Planta* **147**, 495–520.

Franssen, J. M., Cooke, S. A., Digby, J. & Firn, R. D. 1981 Measurements of differential growth causing phototropic curvature of coleoptiles and hypocotyls. *Z. PflPhysiol.* **103**, 207–216.

Franssen, J. M., Firn, R. D. & Digby, J. 1982 The role of the apex in the phototropic curvature of *Avena* coleoptiles: positive curvature under conditions of continuous illumination. *Planta* **155**, 281–286.

Gillespie, B. & Briggs, W. R. 1961 Mediation of geotropic response by lateral transport of auxin. *Pl. Physiol.* **36**, 364–367.

Gillespie, B. & Thimann, K. V. 1963 Transport and distribution of auxin during tropistic responses. I. The lateral migration of auxin in geotropism. *Pl. Physiol.* **38**, 214–225.

Goldsmith, M. H. M., Caubergs, R. J. & Briggs, W. R. 1980 Light-inducible cytochrome reduction in membrane preparations from corn coleoptiles. I. Stabilization and spectral characterization of the reaction. *Pl. Physiol.* **66**, 1067–1073.

Heelis, P. F., Parsons, B. J., Phillips, G. O. & McKellar, J. F. 1978 A laser flash photolysis study of the nature of flavin mononucleotide triplet states and the reactions of the neutral form with acids. *Photochem. Photobiol.* **28**, 169–173.

Hemmerich, P., Massey, V. & Weber, G. 1967 Photo-induced benzyl substitution of flavins by phenylacetate: a possible model for flavoprotein catalysis. *Nature, Lond.* **213**, 728–730.

Iino, M. 1982*a* Action of red light on indole-3-acetic-acid status and growth of coleoptiles of etiolated maize seedlings. *Planta* **156**, 21–32.

Iino, M. 1982*b* Inhibitory action of red light on the growth of the maize mesocotyl: evaluation of the auxin hypothesis. *Planta* **156**, 388–395.

Iino, M. & Carr, D. J. 1981 Safelight for photomorphogenetic studies: infrared radiation and infrared-scope. *Pl. Sci. Lett.* **23**, 263–268.

Jayaram, M., Presti, D. & Delbrück, M. 1979 Light-induced carotene synthesis in *Phycomyces*. *Expl Mycol.* **3**, 42–52.

Kamin, H., Masters, B. S. S. & Gibson, Q. H. 1966 NADPH-cytochrome *c* oxidoreductase. In *Flavins and flavoproteins* (ed. E. C. Slater), pp. 306–324. Amsterdam, London and New York: Elsevier.

Lang-Feulner, J. & Rau, W. 1975 Redox dyes as artificial photoreceptors in light-dependent carotenoid synthesis. *Photochem. Photobiol.* **21**, 179–183.

Leach, C. M. 1968 An action spectrum for light inhibition of the 'terminal phase' of photosporogenesis in the fungus *Stemphylium botryosum*. *Mycologia* **60**, 532–546.

Lenci, F. & Columbetti, G. 1978 Photobehaviors of microorganisms. A biophysical approach. *A. Rev. Biophys. Bioengng* **7**, 341–361.

Leong, T.-Y. & Briggs, W. R. 1981 Partial purification and characterization of a blue light-sensitive cytochrome–flavin complex from corn membranes. *Pl. Physiol.* **67**, 1042–1046.

Leong, T.-Y. & Briggs, W. R. 1982 Evidence from studies with acifluorfen for participation of a flavin–cytochrome complex in blue light photoreception for phototropism of oat coleoptiles. *Pl. Physiol.* **70**, 875–881.

Leong, T.-Y., Vierstra, R. D. & Briggs, W. R. 1981 A blue light-sensitive cytochrome-flavin complex from corn coleoptiles. Further characterization. *Photochem. Photobiol.* **34**, 697–703.

Mandoli, D. F. & Briggs, W. R. 1981 Phytochrome control of two low irradiance responses in etiolated oat seedlings. *Pl. Physiol.* **67**, 732–739.

Mikolajaczyk, E. & Diehn, B. 1975 The effect of potassium iodide on photophobic responses in *Euglena*: evidence for two photoreceptor pigments. *Photochem. Photobiol.* **22**, 269–271.

Muñoz, V. & Butler, W. L. 1975 Photoreceptor pigment for blue light in *Neurospora crassa*. *Pl. Physiol.* **55**, 421–426.

Paietta, J. & Sargent, M. L. 1981 Photoreception in *Neurospora crassa*: correlation of reduced light sensitivity with flavin deficiency. *Proc. natn. Acad. Sci. U.S.A.* **78**, 5573–5577.

Pickard, B. G. & Thimann, K. V. 1964 Transport and distribution of auxin during tropic response. II. The lateral migration of auxin during phototropism of coleoptiles. *Pl. Physiol.* **39**, 341–350.

Pratt, L. H. & Briggs, W. R. 1966 Photochemical and non-photochemical reactions of phytochrome *in vivo*. *Pl. Physiol.* **41**, 367–374.

Quail, P. H. 1979 Plant cell fractionation. *A. Rev. Pl. Physiol.* **30**, 425–484.

Rau, W. 1967 Untersuchungen über die lichtabhängige Carotenoidsynthese. I. Das Wirkungsspektrum von *Fusarium aquaeductuum*. *Planta* **72**, 14–28.

Sargent, M. L. & Briggs, W. R. 1967 The effects of light on a circadian rhythm of conidiation in *Neurospora*. *Pl. Physiol.* **42**, 1504–1510.

Schmidt, W., Hart, J., Filner, P. & Poff, K. L. 1977 Specific inhibition of phototropism in corn seedlings. *Pl. Physiol.* **60**, 736–738.

Senger, H. (ed.) 1980 *The blue light syndrome*. (665 pages.) Berlin, Heidelberg and New York: Springer-Verlag.

Senger, H. 1982 The effect of blue light on plants and microorganisms. *Photochem. Photobiol.* **35**, 911–920.

Senger, H. & Briggs, W. R. 1981 The blue light receptor(s): primary reactions and subsequent metabolic changes. In *Photochemical and photobiological reviews* (ed. K. Smith), vol. 6, pp. 1–38. New York: Plenum.

Shropshire, W. Jr 1980 Carotenoids as primary photoreceptors in blue-light responses. In *The blue light syndrome* (ed. H. Senger), pp. 172–186. Berlin, Heidelberg and New York: Springer-Verlag.

Shropshire, W. Jr & Withrow, R. B. 1958 Action spectrum of phototropic tip-curvature of *Avena*. *Pl. Physiol.* **33**, 360–365.

Tan, K. K. 1978 Light-induced fungal development. In *The filamentous fungi* (ed. J. E. Smith & D. R. Berry), vol. 3 (*Developmental mycology*), pp. 334–357.

Thimann, K. V. & Curry, G. M. 1960 Phototropism and phototaxis. In *Comparative biochemistry* (ed. H. S. Mason & M. Florkin), vol. 1, pp. 243–309. New York: Academic Press.

Went, F. W. & Thimann, K. V. 1937 *Phytohormones*. (294 pages.) New York: Macmillan.

Zalokar, M. 1955 Biosynthesis of carotenoids in *Neurospora*. Action spectrum of photoactivation. *Archs Biochem. Biophys.* **56**, 318–333.

Zimmerman, B. K. & Briggs, W. R. 1963 Phototropic dosage–response curves for oat coleoptiles. *Pl. Physiol.* **38**, 237–247.

Discussion

A. W. GALSTON (*Plant Breeding Institute, Trumpington, U.K.*). Dr Briggs has proposed, on the basis of slightly different action spectra, that different photoreceptors exist for several of the blue-light-mediated reactions in plants. Because it is well known that the same chromophore attached to different proteins can yield slightly different absorption spectra, would Dr Briggs consider that this type of situation could satisfactorily explain the slightly differing flavin-like action spectra? In other words, could these be the 'different' photoreceptors that he calls for?

W. R. BRIGGS. There is no reason why the 'different' photoreceptors I am suggesting could not each be flavins in different environments: on different proteins, in different membranes, etc.

A. W. GALSTON. Is anything known about the mechanism of the light-fleck-induced leaf movements in *Oxalis*? Is there, for example, a trans-pulvinar shuttle of ions, as in *Albizzia* and *Samanea*?

W. R. BRIGGS. To my knowledge, there has been no work at all on the mechanism of *Oxalis* leaf movement, though the elegant work that Dr Galston and Ruth Satter and colleagues have done on *Albizzia* and *Samanea* certainly suggests things to investigate. Indeed, the pulvini are large, the plants available in large numbers (and easy to grow in the greenhouse), and *Oxalis oregana* might be first-rate for studying some aspects of leaf movement.

J. DIGBY AND R. D. FIRN (*Department of Biology, University of York, U.K.*). In contrast to the results that Dr Briggs has presented, Mr K. MacLeod in our laboratory has recently found that in dark-grown *Avena* coleoptiles, first positive phototropic curvature is brought about almost entirely by a growth retardation on the illuminated side, and the growth rate of the shaded side is virtually unchanged. It is possible, as in the phototropism of some hypocotyls, that the contribution to curvature of changes in growth rate on the shaded side may vary, depending on the growth conditions. This suggests that simple curvature requirements are really inadequate as a means of describing phototropism, yet it is just such simple curvature measurements that we seem to rely on for phototropic action spectra.

W. R. BRIGGS. There is certainly no question that under different conditions the relative contribution of growth rate changes on the two sides may vary widely. However, simply because both sides of a unilaterally irradiated coleoptile slow down, one more than the other (for example) does not in itself eliminate a lateral auxin differential in the Went–Cholodny sense from being the reason for curvature. If there are two separate photoreactions, one simply a general inhibition of growth, and the other an induction of lateral transport of auxin, it would still be the lateral transport of auxin that brought about the curvature response. The questioners are quite correct in that if this is so (multiple photosystems and multiple light responses), then action spectroscopy is not as helpful as one might hope. In the present limited experiments, we have simply tried to saturate any red-induced effects on growth, as a way of isolating blue-light effects on phototropism. The results *in this case* are clearly consistent with the Went–Cholodny hypothesis. Until similar kinds of experiments with other curvature responses are demonstrated, we do not extrapolate beyond this one system, first positive curvature in corn, grown under these specific conditions.

Phil. Trans. R. Soc. Lond. B **303**, 361–375 (1983)

Printed in Great Britain

Molecular properties of phytochrome

BY M. FURUYA

Biology Department, Faculty of Science, University of Tokyo, Hongo, Tokyo 113, *Japan
and Division of Biological Regulation, National Institute for Basic Biology, Okazaki* 444, *Japan*

Chromopeptides with molecular masses of *ca.* 114, 62, 56, 40, 39 and 33 kDa were prepared from pea phytochrome by limited proteolysis. Absorption and circular dichroism spectra were determined and proton uptake and release investigated. The data indicate how long the chromopeptide chain must be for photoreversible changes between P_r and P_{fr} or between P_{659} and P_{bl}.

Double flash-photolytic and low-temperature spectroscopic studies on the phototransformation pathways from P_r to P_{fr} and from P_{fr} to P_r of native and degraded chromopeptides were carried out under different conditions, demonstrating that the pool size of kinetically detectable intermediates in a sample changed reversibly depending upon monomer size, and microenvironmental factors such as pH and temperature.

Six monoclonal antibodies against rye phytochrome and six against pea phytochrome were raised and investigated in terms of the sites of phytochrome determinants, species specificity, and influence on spectral and other molecular properties.

1. INTRODUCTION

Developmental and physiological processes in plants are controlled by not only genetic information but also by changes in the physical and chemical factors of the environment. Among the latter, light is well known to be the most evident and crucial factor in plants (Smith 1976, 1982). Phytochrome and blue–near-u.v. light-absorbing pigments are widely distributed in the plant kingdom as phototransducers and act for a variety of photomorphogenetic responses (Furuya 1968; Senger 1980; Shropshire & Mohr 1983), and each developmental process such as dormancy induction, dormancy breakage, cell division, growth and differentiation is properly progressed under a collaboration of these two photoreceptor systems (Furuya 1978, 1980).

Phytochrome is a chromoprotein with two distinct and photointerconvertible forms, a form absorbing red light, P_r, and a form absorbing far-red light, P_{fr} (Butler *et al.* 1959). Phytochrome can be isolated from plant tissues and purified by conventional and affinity procedures, so spectrophotometrical and molecular properties of phytochrome *in vitro* have been intensively studied in the past two decades, and the results accumulated in the literature have been repeatedly reviewed by Pratt (1978, 1979, 1982*a*, *b*). I therefore do not intent to present a general review of this subject, but should like to introduce here some recent results obtained in my laboratories.

2. SUBUNIT SIZE AND SPECTRAL PROPERTIES

Since the discovery of phytochrome (Butler *et al.* 1959), photoreversible absorbance changes between P_r and P_{fr} have been the most prominent property of phytochrome. Although absorption spectra *in vivo* result from all the pigments existing in the sample tissues (figure 1*a*),

[15]

difference spectra after actinic red and far-red light irradiations clearly separate phytochrome from other pigments (figure 1*b*). Thus the detection of phytochrome both *in vivo* and *in vitro* has been most widely based upon this spectral property.

Phytochrome has been isolated from various plant tissues and purified by either conventional or immunoaffinity procedures (see Pratt 1982*a*). In the early days of phytochrome study, the

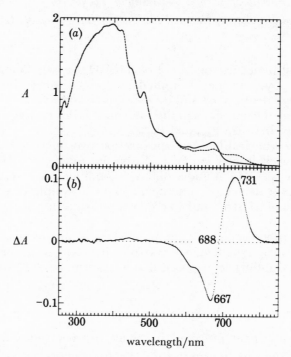

FIGURE 1. Absorption (*a*) and difference (*b*) spectra of etiolated pea hook tissue (Y. Inoue, unpublished data). ——, Totally dark-grown tissue; ---, after irradiation with red light.

monomer size of phytochrome was reportedly believed to be *ca.* 60 kDa (Mumford & Jenner 1966). However, phytochrome was later found to consist of two identical chromopeptides, each of molecular mass *ca.* 120 kDa (Briggs & Rice 1972). By now it is well established that the former, 'small' phytochrome, is produced from the latter, 'large' phytochrome, by proteolytic degradation (Pratt 1982*a*). It has, however, long been questioned why the absorption maximum of purified 'large' phytochrome in the literature (*ca.* 724–725 nm; figure 3*a*) was somewhat, but significantly, shorter than that observed *in vivo* (730–732 nm; figure 1). Recently it became evident that phytochrome isolated from oats as P_{fr} absorbs at longer wavelengths than that extracted as P_r and that the long-wavelength absorbing phytochrome apparently shows a larger molecular mass on gel electrophoresis than that reported as 'large' phytochrome (Epel 1981; Baron & Epel 1982). Further, Vierstra & Quail (1982*a*, *b*) demonstrated that 'native' oat phytochrome is homogeneous, with a monomeric molecular mass of 124 kDa, whereas 'large' phytochrome purified by conventional procedure is heterogeneous with molecular masses of 118, 114 and 112 kDa. It appears that the 'native' phytochrome can be prepared whenever endogenous proteolysis is properly prevented during extraction and purification. An example of rye phytochrome is presented in figure 2, which confirms the above conclusion. An immunoaffinity purification procedure provides a major band of

124 kDa monomer at the final purification step irrespective of the spectral form in which phytochrome is extracted (K. T. Yamamoto, unpublished).

The spectral properties of 'small' phytochrome are generally very similar to those of 'large' phytochrome (Pratt 1978, 1979). The next step is to characterize the autonomous chromophore-containing subregion of phytochrome (chromophore domain) that can exhibit the

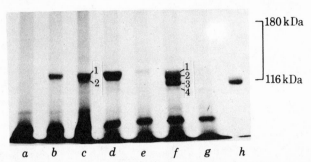

FIGURE 2. Sodium dodecyl sulphate polyacrylamide gel electrophoretograms of immunoprecipitated rye phytochrome (K. T. Yamamoto, unpublished). Phytochromes immunoprecipitated from crude extracts as P_{fr} (a, b) or P_r (c) were compared with those immunoprecipitated from purified phytochrome preparations obtained by conventional procedures as P_{fr} (d, e) or P_r (f, g). Conventionally purified pea phytochrome was also included in the electrophoresis for comparison (h). Immunoprecipitations were performed with monoclonal anti-rye phytochrome antibody (AR3)-coated *Staphylococcus aureus* cells (b, c, d, f) according to the method of Vierstra & Quail (1982a). A monoclonal antibody to rat liver cell membrane glycoproteins (Fukumoto *et al.* in preparation) was also used as a control for non-specific binding of antibodies (a, e, g). The immunoprecipitates were subjected to sodium dodecyl sulphate polyacrylamide electrophoresis (Laemmli 1970) using 30 g l⁻¹ acrylamide stacking gel and 50 g l⁻¹ separating gel.

photoreversible absorbance change by itself. We have prepared five chromophore-containing fragments from pea phytochrome of 114 kDa in the P_r form by limited proteolysis with trypsin, thermolysin and chymotrypsin, and determined the absorption and circular dichroism (c.d.) spectra, and the fluences required for photoconversion (Yamamoto & Furuya 1983). The fragments of 62 and 56 kDa that were 'small' phytochrome (figure 3b) showed a photoreversible transformation between P_{667} and P_{722} like 'large' phytochrome (figure 3a). Smaller fragments of 40, 39 and 33 kDa showed an absorption maximum at 657–660 nm (P_{658}), which was transformed to a bleached form (P_{bl}) after a brief exposure to red light (figure 3c). P_{bl} was transformed back to P_{658} by far-red light at a fluence that was *ca.* 10 times that needed for the conversion of P_{722} to P_{667}. The transformation between P_{658} and P_{bl} was repeatedly photoreversible. Both P_{658} and P_{bl} showed negative c.d. bands in the red region like P_{667}, whereas P_{722} has a positive band in the far-red region. As far as the size of the chromophore domain of phytochrome is concerned, the smallest fragment that showed photoreversible transformation between P_r and P_{fr} was a 56 kDa chromopeptide obtained by thermolysin digestion, and the largest one that no longer exhibited the P_r–P_{fr} photoconversion was a 40 kDa fragment prepared by trypsin digestion. This fact indicates that the molecular mass of the chromophore domain of phytochrome is at most 56 kDa. This conclusion is consistent with that in a previous report (Stoker *et al.* 1978).

The fragments of 40 kDa or smaller obtained by limited trypsin digestion did not show the typical photoreversible conversion between P_r and P_{fr}, but exhibited an atypical photoreversible conversion between new spectral forms, P_{658} and P_{bl}, by actinic red and far-red light. The fluence response of the conversion showed that incomplete conversion was due to photoequilibrium between P_{658} and P_{bl}. These facts show that digestion of the 56 kDa fragment to the 40 kDa fragment distorts the confrontation around its chromophore, but that the 40 kDa fragment still maintains a certain structure that is essential for the photoreversible spectral change of phytochrome.

It is interesting to note that the absorption spectrum of P_{bl} with 33–40 kDa fragments

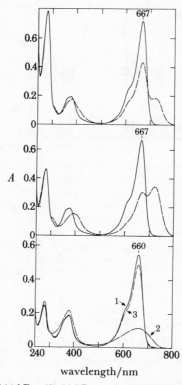

FIGURE 3. Absorption spectra of (*a*) 114 kDa, (*b*) 62 kDa and (*c*) 40 kDa chromophore-containing fragments of pea phytochrome in 0.1 M sodium phosphate (pH 7.8), 1 mM Na_2EDTA and 0.25 mM dithiothreitol (Yamamoto & Furuya 1983). The absorption spectra were determined at 3 °C with a dual-wavelength difference spectrophotometer (Hitachi model 557) with 1 cm light-path quartz cuvettes and a slit width of 1 nm. (*a*), (*b*) Solid line, P_r; broken line, red-light-induced photostationary state. (*c*) Curve 1, an initial spectrum before actinic irradiation; curve 2, a spectrum after saturating red light irradiation; curve 3, a spectrum after saturating far-red light irradiation after the red light.

(Yamamoto & Furuya 1983) was essentially the same as that of phytochrome in the presence of urea (Butler *et al.* 1964), divalent metallic ions (Pratt & Cundiff 1975), anilinonaphthalene sulphonate (ANS) (Hahn & Song 1981), liposomes (Furuya *et al.* 1981), and a triterpenoid saponin (Konomi *et al.* 1982), although phytochrome did not show the repeated photoconversion between P_{659} and P_{bl} in the presence of these substances, with the exception of the saponin. In fact, P_{bl} was formed as a result of the interaction of the saponin with one or more intermediates produced during the phototransformation of P_r to P_{fr} (Konomi *et al.* 1982), because the addition of the saponin to phytochrome after red light irradiation did not produce P_{bl} as effectively as did red light irradiation of P_{658}, which was formed immediately after the saponin was added to P_{667}. For ANS it was postulated that phototransformation intermediates of phytochrome were selectively complexed with ANS, resulting in P_{bl} (Hahn & Song 1981). In addition, phytochrome in the presence of ANS showed a stronger bleaching of P_r, a significant loss of photoreversibility and much smaller negative c.d. band of the bleached form, indicating that the spectral forms observed in the presence of ANS were distinct from those observed in the 33–40 kDa fragments.

In conclusion, only three patterns of photoreversible spectral changes of phytochrome (table 1) have been found under experimental conditions so far tested: namely (1) phototransformations between P_{660} and P_{730} in the 'native' form; (2) that between P_{666} and P_{725} in 'large'

TABLE 1. MONOMER SIZE AND SPECTRAL PROPERTIES OF PHYTOCHROME IN VARIOUS PLANTS

sample	monomer molecular mass/kDa	P_r max.	difference spectra/nm isosbestic point	P_{fr} max.	reference
in vivo					
oat (coleoptile)	.	666	687	729	
(mesocotyl)	.	668	689	730	
corn (mesocotyl)	.	669	690	732	Inoue (unpublished)
rye (coleoptile)	.	665	688	733	
pea (hook)	.	667	688	730	
in vitro					
'native'					
oat	124	660	.	730	Vierstra & Quail (1982)
rye	124	662	684	731	Inoue & Yamamoto (unpublished)
'large'					
oat	118, 114	665	.	722	Vierstra & Quail (1982)
rye	124–118	665	687	729	Inoue & Yamamoto (unpublished)
pea	114	667	.	722	Yamamoto & Furuya (1983)
'small'					
oat	62	666	.	725	Rice *et al.* (1973)
rye		665	687	726	Inoue (unpublished)
pea	62	667	—	722	Yamamoto & Furuya (1983)
degraded					
pea	33–40	658	.	bleached	Yamamoto & Furuya (1983)

and 'small' phytochrome; and (3) that between P_{658} and P_{bl} in 33–50 kDa chromopeptide fragments and in 'small' or 'large' phytochromes in the presence of the above chemicals. Therefore, the phytochrome chromophore appears to take only a few quasi-stable spectral states under various microenvironment rather than a great number of different states, which may result in diverse spectra.

3. PHOTOREVERSIBLE PROTON UPTAKE AND RELEASE

Proton uptake and release have been proposed in hypothetical models for phytochrome phototransformation (Rüdiger 1980; Lagarias & Rapoport 1980). In this connection, we recently reported the first crucial evidence of photoreversible proton transfer in unbuffered solutions of both 'large' and 'small' pea phytochromes (Tokutomi *et al.* 1982). When pH changes of dialysed phytochrome solutions were measured with a semimicro-combination pH electrode, red light irradiation caused an alkalinization of the solutions in the pH range 5.2–7.6 and an acidification in the pH range 7.6–9.2 (Tokutomi *et al.* 1983a). The pH changes were fully reversed by a subsequent irradiation with far-red light, and the effects of red and far-red light were repeatedly photoreversible.

These photoreversible pH changes of unbuffered pea phytochrome solutions occurred concomitantly with not only optically measured absorbance changes between P_r and P_{fr} that were induced by actinic light, but also in dark reversion of P_{fr} to P_r. The net numbers of

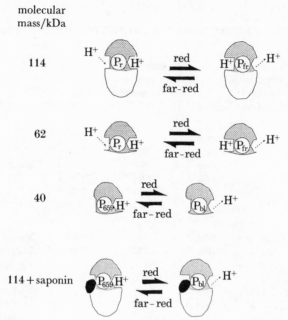

FIGURE 4. Schematic illustration of proton uptake and release in unbuffered solution of pea phytochrome. For details, see the text.

protons released from, or taken up to, phytochrome were dependent upon the degree of absorbance change at 730 nm. The alkalinization in the dark took place slowly in parallel with P_{fr} dark reversion, whereas the acidification in the dark showed an initial rapid phase correlated to the rapid absorbance increase at 667 nm (Tokutomi *et al.* 1983*a*). This finding is consistent with the observation of Sarkar & Song (1981) of the effect of D_2O on phototransformation and dark transformation of oat phytochrome.

Unbuffered solution of pea 'large' phytochrome in the pH range 7.4–7.8 was transiently acidified at an early step of the process after irradiation with red light. The pH titration curves of the solutions with both P_r and a photostationary state under red light irradiation were determined, and the resultant curve of the latter showed an upward shift at pH 5.4–7.6 and a downward shift at pH 7.6–9.2. The results on the transient acidification and the light-induced shift of pH titration curves suggest that each phytochrome molecule has one or more proton release sites and one or more uptake sites, as shown in figure 4, and that the former possibly has one or more groups with a pK of *ca.* 6.0 and undergoes an upward pK shift, while the latter may have one or more groups with a pK of 9.2 or higher and undergoes a downward shift (Tokutomi *et al.* 1983*a*).

Finally, proton uptake or release, if any, during phototransformation between P_{659} and P_{bl}, which were described in the previous section (see figure 3*c*), were measured. When an unbuffered solution of proteolytically prepared 40 kDa chromopeptide of pea phytochrome was irradiated with red light, an acidification of the solution was induced at pH 7.0–8.7 as 'large' pea phytochrome, but no alkalinization was observed at any pH range tested. Similarly, an acidification (but no alkalinization) was observed in 'large' pea phytochrome solution in the presence of 0.8 mM soyasaponin I (Tokutomi *et al.* 1983*b*), indicating that the phototransformation between P_{659} and P_{bl} results in the acidification but not in the alkalinization.

4. PHOTOTRANSFORMATION PATHWAYS AND INTERMEDIATES

Photoreversible absorbance changes of phytochrome result from a series of physical and chemical changes in the phytochrome molecule that are induced by photon capture by the chromophore and its intramolecular effects on protein moiety. Although P_r and P_{fr} are spectrophotometrically detectable as stable forms at physiological temperature, numerous short-lived intermediates of phytochrome have been characterized in both directions of phototransformation *in vivo* and *in vitro* by flash kinetic spectroscopy and low-temperature spectroscopy.

(a) Flash photolysis

Four intermediates of 'small' oat phytochrome were originally demonstrated to occur through two kinetically identifiable stages by a flash photolysis technique on the pathway from P_r to P_{fr} (Linschitz *et al.* 1966), and two more were added to the later stage (Pratt & Butler 1970). Recently the phototransformation intermediates from P_r to P_{fr} were re-examined with 'large' phytochrome of pea (Shimazaki *et al.* 1980; Cordonnier *et al.* 1981) and oat (Cordonnier *et al.* 1981; Pratt *et al.* 1982), demonstrating that phototransformation pathways from P_r to P_{fr} of 'large' phytochrome took place, as with 'small' phytochrome, through at least three reaction stages. These include a photoinduction of I_{692}, a dark decay process of I_{692} to I_{bl} on a microsecond timescale and the appearance of P_{fr} from I_{bl} on a millisecond or longer timescale. The last two dark processes were separable kinetically into three distinct reactions (figure 5). Flash activation of intermediates present during the first decay stages resulted in their photoconversion back to P_r within 8 µs, and during the second stage converted them to another transient intermediate stage, which decayed thermally to P_r within 2 ms (Pratt *et al.* 1983).

In contrast, significantly less attention has so far been paid to the phototransformation of P_{fr} to P_r (e.g. 'small' oat phytochrome (Linschitz *et al.* 1966; Pratt & Butler 1970) and 'large' pea phytochrome (Inoue *et al.* 1982)). One of the difficulties for such a study is due to the dark

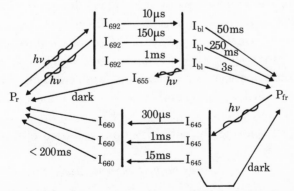

FIGURE 5. A general scheme showing probable phototransformation pathways and intermediates of phytochrome. The data are taken from Shimazaki *et al.* (1980), Inoue *et al.* (1982) and Pratt *et al.* (1983).

reversion of P_{fr} to P_r, which occurs in parallel with the phototransformation. When P_{fr} of 'large' pea phytochrome was excited by a 715 nm laser flash light (Inoue *et al.* 1982), the maximum amount of phototransformation intermediates was produced by a pulse of 50 mJ, which resulted in *ca.* 65% P_r at the photostationary state. A difference spectrum between an intermediate measured 10 μs after flash excitation and P_{fr} showed an absorbance increase at 651 nm and a decrease at 724 nm, indicating a formation of the first detectable intermediate (I_{645} in figure 5). The difference spectrum of this earliest intermediate showed a peak at 651 nm, which was similar to the previously reported data on 'small' oat phytochrome (Linschitz *et al.* 1966; Pratt & Butler 1970). Although Kendrick & Spruit (1977) found an intermediate named *lumi*-F *in vivo* and *in vitro*, the difference spectrum of 'large' pea phytochrome did not show any positive peak at 730 nm, so that the existence of *lumi*-F seems dubious in this sample. The observation that the absorbance of 'large' pea phytochrome increased in both red and far-red spectral regions in the dark after an excitation with a far-red laser flash light (Inoue *et al.* 1982) indicates that some of the induced intermediates were transformed back to P_{fr} besides the phototransformation from P_{fr} to P_r (figure 5). Considering this dark process, the decay curve of I_{645} was determined at 554 nm, at which the absorbance of the intermediate rapidly decreased. The resultant curve was kinetically resolved into three reactions with rate constants of 2500, 590 and 48 s^{-1}.

Spectral changes in the visible and near-u.v. region during phototransformations from P_r to P_{fr} (Shimazaki *et al.* 1980) and from P_{fr} to P_r (Inoue *et al.* 1982) became available with the use of a custom-built multichannel transient spectral analyser that can determine the absorbance change in the spectral range between 350 and 800 nm at once with a sampling time of 300 μs beginning 10 μs after a flash excitation (Furuya *et al.* 1983). Difference spectra between phototransformation intermediates and P_r showed that a small but significant increase at 400–410 nm and decrease at 360 nm took place 10–260 μs after a red light laser flash, and those between intermediates and P_{fr} indicated that an increase in absorbance at 370–380 nm and a decrease around 415 nm occurred 10–310 μs after a far-red flash. Using these data on intermediate spectra, Sugimoto *et al.* (1983) analysed theoretically the change of chromophore structure of phytochrome during the phototransformations in terms of wavelength and oscillator strength of absorption by using the zero-differential approximation of molecular orbital theory for π-electrons. The effects of a point charge and a point dipole on the optical absorption of phytochromobilin intermediates were examined by the stationary perturbation theory for degenerate states. The results indicate that the *cis–trans* photoisomerization of pyrrole ring D, if any, occurred within 10 μs after a laser-flash excitation of phytochrome and that the conformation of phytochromobilin and the protein moiety did not significantly change during the examined period of phototransformations in both directions.

The results on phototransformation pathways of 'large' pea and oat phytochrome obtained by laser flash photolysis in my laboratory are summarized in figure 5. The three elementary reaction stages in each direction were kinetically separatable in both directions of the phototransformation pathway: namely, I_{692}, I_{bl} and P_{fr} were found to be formed on the way from P_r to P_{fr} (Pratt *et al.* 1982), and similarly I_{645}, I_{660} and P_r were involved on the way from P_{fr} to P_r (Inoue *et al.* 1982). Each stage in both directions consisted of three kinetically distinct reactions (figure 5).

(b) Heterogeneity of monomer size

A question thus arises as to whether the triple pathways in each stage result from subunit heterogeneity of the samples, such as the 124, 118 and 114 kDa monomers (table 1), or whether it happens even with a sample of homogeneous monomer size. To answer this question, we have carefully extracted homogeneous samples of 124 and 62 kDa rye phytochrome (figure 2), and re-examined the phototransformation pathways. The results clearly showed that three kinetically distinguishable reactions were separated with both homogeneous samples, and that when the mixture of 124–118 kDa samples was measured, we separated more than four stages of P_{fr} appearance from I_{bl} (Inoue, Yamamoto & Furuya, unpublished).

(c) Low-temperature spectroscopy

The phototransformation of P_r to P_{fr} in 'small' oat phytochrome was mainly studied by low-temperature spectroscopy (Spruit 1975), and that in 'large' oat phytochrome was recently reported by Song et al. (1981). We have therefore intensively investigated the spectral changes accompanying phototransformation in both directions by low-temperature spectroscopy with 'large' pea phytochrome (Sasaki et al. 1983 a, b), using the same samples as in the above studies on laser-flash photolysis (Shimazaki et al. 1980; Inoue et al. 1982; Pratt et al. 1983). It is fruitful to compare the data obtained by these two different techniques. It became evident that an irradiation of P_r with red light below $-80\,^{\circ}\mathrm{C}$ yielded an intermediate, I_{693}. Part of this intermediate reverted to P_r if kept above $-130\,^{\circ}\mathrm{C}$, and the residue was converted to a 'bleaching form (I_{bl})' with two components. On warming, it was finally converted to P_{fr} at $-40\,^{\circ}\mathrm{C}$ or higher. In the phototransformation of P_{fr} to P_r an intermediate (I_{670}) was formed below $-80\,^{\circ}\mathrm{C}$, which was converted by warming above $-80\,^{\circ}\mathrm{C}$ to I_{665}, with two components. The intermediates were then transformed to P_r via I_{665}^{*}. The elementary pathways separated in this work are shown schematically in figure 6.

It cannot yet be decided whether the courses of the conversion from I_{693} to P_{fr} and from I_{670} to P_r are parallel or sequential. However, we found many similarities between experimental results obtained by flash photolysis near room temperature and those by low-temperature spectrophotometry. One of the typical examples is that the spectral changes in the conversion of I_{693} were similar to those observed by Shimazaki et al. (1980). As already described, the formation of I_{693} was dependent on temperature, owing to the dark reversion of I_{693}-I to P_r above $-130\,^{\circ}\mathrm{C}$. Moreover, because I_{693}-II was converted to I_{bl}, composed of two molecular species, it was concluded that I_{693} should consist of three components.

In the phototransformation of P_{fr} to P_r, the rate of formation of I_{670} decreased as the temperature was lowered, and at least three intermediates (I_{665} (low), I_{665} (high) and I_{665}^{*}) were observed in the course of the thermal conversion of I_{670} to P_r. The large spectral shift due to photoreaction of P_{fr} to I_{670} suggests that I_{670} may have a π-electron system of the chromophore

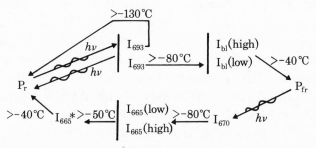

FIGURE 6. A scheme for the photoreaction cycle of phytochrome proposed on the basis of low-temperature spectrophotometry (Sasaki et al. 1983 b). For details see the text.

close to that of P_r. The conversion of I_{670} to P_r in the dark might be induced by conformational changes of the protein moiety surrounding the chromophore.

5. MONOCLONAL ANTIBODIES TO PHYTOCHROME

The molecular properties and function of phytochrome have been immunochemically studied by using multiclonal antibodies (Hopkins & Butler 1970; Rice & Briggs 1973; Pratt 1973; Cordonnier & Pratt 1982). As the conventional antibodies produced by multiple immunochemical determinants showed diverse antigenic specificity, their application in the analysis of phytochrome's molecular structure has been limited, and in particular they cannot be used to identify the entity governing the action of phytochrome. The recently developed method of producing monoclonal antibodies (Köhler & Milstein 1975), however, provides an antibody of undoubted specificity and unlimited supply, and it is possible to produce monoclonal antibodies that are specific to different determinants of the surface of the phytochrome molecule.

We have recently succeeded in producing six monoclonal antibodies (AR1 to AR6) to rye phytochrome (Nagatani *et al.* 1983*a*) and six (AP1 to AP6) to pea phytochrome (Nagatani *et al.* 1983*b*) by fusing spleen cells from immunized BALB/c mice with NS-1 myeloma cells. All the clones have been stably producing each monoclonal antibody in ascitic fluids of pristane-treated BALB/c mice into which the hybridomas were intraperitoneally injected. The binding ability of the monoclonal antibodies (mAbs) to phytochrome, which were coupled to either sheep red blood cells (s.r.b.c.) or Sepharose 4B, were determined by radioimmunoassay. The six mAbs to rye phytochrome had titres (defined as the dilution giving 50% binding in radioimmunoassay) of between $1:8 \times 10^4$ and $1:8 \times 10^5$.

The results of immunoelectrophoresis of these mAbs with the use of rabbit anti-mouse IgG_1 serum indicated that two of the six ARs were of the IgG_2 type and the other four were of the IgG_1 type. It is very important to know the determinant sites of the phytochrome molecule that resulted in each mAb. Thus the reactivities of each mAb to 'small' phytochrome and variously digested chromopeptides were measured by the inhibition assay (Mason & Williams 1980). All six mAbs of rye phytochrome were able to bind to the 40 kDa fragments, although the ratios of the molar concentrations giving 50% inhibition for 'small' phytochrome to that for 'large' phytochrome were different. Namely, the ratios for AR1, AR2, AR3 and AR4 were significantly higher than those for AR5 and AR6, whereas the ratios for AR5 and 6 were lowest, in the neighbourhood of 1.5.

Absorption spectra of P_r, and that at red-light-induced photoequilibrium, were measured after incubation of 1 μM P_r with *ca.* 10 μM crude IgG in phosphate-buffered saline at 4 °C in the dark overnight. None of the six IgGs affected the spectra. This fact that the binding of mAbs to phytochrome does not influence the photoreversible spectral change suggests a possibility that determinants of all the tested mAbs are located in the peptide regions that do not affect the chromophoric domain of phytochrome. Reactivities of the mAbs against 'large' rye phytochrome were determined with P_r in the dark or with phytochrome under continuous red light at 4 °C overnight to check the difference between affinities of the mAbs to P_r and P_{fr}. Their affinities to P_r and P_{fr} were indistinguishable in radioimmunoassay.

The cross-reactivity of the twelve available mAbs toward phytochromes provided from four dicotyledonous and three monocotyledonous species was examined by inhibition assay.

AR1–AR6 did not cross-react with pea phytochrome at the maximum concentration tested (0.4 mg ml^{-1}), whereas AR5 and AR6 did cross-react with 'large' oat phytochrome (table 2). It is interesting to note that AR6 showed a stronger affinity to 'large' oat phytochrome than to 'large' rye phytochrome. The results so far obtained are summarized in table 2. At present, none of the mAbs react with both monocotyledonous and dicotyledonous phytochromes.

TABLE 2. CROSS-REACTIVITY OF THE MONOCLONAL ANTIBODIES OF RYE AND PEA PHYTOCHROMES
TOWARD PHYTOCHROMES PREPARED FROM VARIOUS PLANT SPECIES

phytochrome prepared from	anti-rye phytochrome antibodies						anti-pea phytochrome antibodies					
	AR1	AR2	AR3	AR4	AR5	AR6	AP1	AP2	AP3	AP4	AP5	AP6
rye (*Secale cereale*)	+	+	+	+	+	+	−	−	−	−	−	−
oat (*Avena sativa*)	−	−	−	−	+	+	−	−	−	−	−	−
rice (*Oryza sativa*)	−	−	−	−	−	−	−	−	−	−	−	−
pea (*Pisum sativum*)	−	−	−	−	−	−	+	+	+	+	+	+
mung bean (*Vigna radiata*)	−	−	−	−	−	−	−	−	−	(+)	+	+
radish (*Raphanus sativus*)	−	−	−	−	−	−	−	−	−	−	(+)	+
morning glory (*Pharbitis nil*)	−	−	−	−	−	−	−	−	−	−	−	−

Symbols: +, positive reaction; (+), weak positive reaction; −, negative reaction.

Cundiff & Pratt (1975) reported the presence of antibodies that bound to 'large' phytochrome but not to 'small' phytochrome by double diffusion assay. All the six mAbs to rye phytochrome in this work, however, reacted with both 'large' and 'small' phytochromes. In this connection it must be mentioned that the assay system used for the screening of these six mAbs to rye phytochrome employed phytochrome coupled with s.r.b.c. (Nagatani *et al.* 1983*a*). Considering that phytochrome is a multifunctional protein with a chromophoric and a hydrophobic domain and that the hydrophobic domain responsible for the photoinduced increase of hydrophobicity is separated by trypsin digestion for the chromophoric domain, accounts for the phototransformation between P_r and P_{fr} (Tokutomi *et al.* 1981). Although the mechanism of the binding of proteins with s.r.b.cs is poorly understood (Goding 1976), it is possible that phytochrome used in the assay binds to s.r.b.cs through hydrophobic interaction so that the hydrophobic region of phytochrome is hidden by the surface of the s.r.b.cs. In fact, when we used Sepharose 4B instead of s.r.b.cs, we obtained mAbs that showed a greater variety of binding site (Nagatani *et al.* 1983*b*).

6. CONCLUDING REMARKS

As phytochrome-mediated photoreversible responses in plants must result from the difference of molecular properties between P_r and P_{fr}, extensive attempts have been made to find such differences. Many studies, however, failed to find such differences on physical and chemical nature of phytochrome (table 3). Recently, some molecular properties of either the chromophore or the apoprotein have been demonstrated to show differences between P_r and P_{fr} (table 4). This information should be a great help in understanding the primary action of phytochrome, but it is also true that no one yet knows how phytochrome induces photomorphogenetic reactions.

Molecular properties of phytochrome have been studied mainly with 'large' and 'small'

TABLE 3. MOLECULAR PROPERTIES OF PHYTOCHROME THAT SHOWED
NO DIFFERENCES BETWEEN P_r AND P_{fr}

property tested	reference
c.d. spectra in u.v. region	Tobin & Briggs (1973)
n.m.r. spectra of aliphatic protons	Song et al. (1982)
exposed tyrosine and carboxyl groups	Hunt & Pratt (1981)
electrophoretic mobility	Briggs et al. (1968)
surface charges (isoelectric focusing)	Hunt & Pratt (1981)
sedimentation velocity	Briggs et al. (1968)
elution profiles of gel exclusion and brushite chromatography	Briggs et al. (1968)
affinity for anti-'small' phytochrome sera	Pratt (1973) Rice & Briggs (1973)
affinity for anti-'large' phytochrome sera	Cundiff & Pratt (1975)
affinity for monoclonal anti-'large' phytochrome antibodies	Nagatani et al. (1983a)

TABLE 4. MOLECULAR PROPERTIES OF PHYTOCHROME THAT SHOWED
DIFFERENCES BETWEEN P_r AND P_{fr}

property tested	P_r	P_{fr}	reference
spectral lability in the presence of urea	<		Butler et al. (1964)
spectral lability in the presence of Me^{2+}	<		Pratt & Cundiff (1975)
spectral lability in the presence of $KMnO_4$	<		Hahn et al. (1980)
exposed cysteine and histidine	<		Hunt & Pratt (1981)
exposed tryptophan	<		Sarkar & Song (1982)
n.m.r. spectra of aromatic and −NH-proton resonance region	<		Song et al. (1982)
exchangeable proton	<		Hahn & Song (1982)
proton uptake and release	<		Tokutomi et al. (1982)
affinity for alkyl groups	<		Yamamoto & Smith (1981) Tokutomi et al. (1981)
affinity for Cibacron blue F3GA	<		Smith (1981)
affinity for ANS	<		Hahn & Song (1981)
affinity for liposomes	<		Kim & Song (1981)
energy transfer from FMN to phytochrome	+	−	Song et al. (1981)

phytochrome in the past. But, although it seems likely that 124 kDa monomers are 'native' phytochrome, many questions remain unsolved as discussed by Quail (this symposium). As we now can routinely provide 'native' phytochrome, it would be important to re-examine the molecular properties of phytochrome. Furthermore, we really need to know the differences between monomers, dimers and other polymers of phytochrome, if any, with their physical and chemical properties, and it is especially important to demonstrate whether or not phytochrome in cells exists as a dimer, as it does in solution.

Comparing the knowledge that we had a decade ago, we can say that the progress has been enormous, but it is very important to continue investigating the molecular properties of phytochrome. However, it would be more crucial to discover the partner substance(s) that transduce the signal from a phytochrome molecule to cell membranes. In this connection the evidence for the different dichroic orientation of P_r and P_{fr} at cell membranes in *Mougeotia* (Haupt et al. 1969) and *Adiantum* (Kadota et al. 1982) encourages us to work in such direction.

The research on the molecular properties of phytochrome in my laboratories was supported in part by Grants-in-Aid for special research on photophysiology in 1978–81 and for group research on phytochrome in 1981–3. This work has been made possible by collaboration with many experts in various fields, to whom I should like to make grateful acknowledgement of their help. I am specially indebted to Dr Y. Inoue, Dr K. T. Yamamoto and Dr S. Tokutomi for their thorough and painstaking efforts during this work.

References

Baron, O. & Epel, B. L. 1982 Studies on the capacity of P_r *in vitro* to photoconvert to the long-wavelength P_{fr}-form. A survey of ten plant species. *Photochem. Photobiol.* **36**, 79–82.

Briggs, W. R. & Rice, H. V. 1972 Phytochrome: chemical and physical properties and mechanism of action. *A. Rev. Pl. Physiol.* **23**, 293–334.

Briggs, W. R., Zollinger, W. D. & Platz, B. B. 1968 Some properties of phytochrome isolated from dark-grown oat seedlings (*Avena sativa* L.) *Pl. Physiol.* **43**, 1239–1243.

Butler, W. L., Norris, K. H., Siegelman, H. W. & Hendricks, S. B. 1959 Detection, assay, and preliminary purification of the pigment controlling photoresponsive development of plants. *Proc. natn. Acad. Sci. U.S.A.* **45**, 1703–1708.

Butler, W. L., Siegelman, H. W. & Miller, C. O. 1964 Denaturation of phytochrome. *Biochemistry, Wash.* **3**, 851–857.

Cordonnier, M. M., Mathis, P. & Pratt, L. H. 1981 Phototransformation kinetics of undegraded oat and pea phytochrome initiated by laser flash excitation of the red-absorbing form. *Photochem. Photobiol.* **34**, 733–740.

Cordonnier, M. M. & Pratt, L. H. 1982 Comparative phytochrome immunochemistry as assayed by antisera against both monocotyledonous and dicotyledonous phytochrome. *Pl. Physiol.* **70**, 912–916.

Cundiff, S. C. & Pratt, L. H. 1975 Phytochrome characterization by rabbit antiserum against high molecular weight phytochrome. *Pl. Physiol.* **55**, 207–211.

Epel, B. L. 1981 A partial characterization of the long wavelength 'activated' far-red absorbing form of phytochrome. *Planta* **151**, 1–5.

Furuya, M. 1968 Biochemistry and physiology of phytochrome. *Progr. Biochem.* **1**, 347–405.

Furuya, M. 1978 Photocontrol of developmental processes in fern gametophytes. *Bot. Mag., Tokyo* (special issue) **1**, 219–242.

Furuya, M., Inoue, Y. & Maeda, Y. 1983 Design and performance of a multichannel transient spectra analyzer. *Photochem. Photobiol.* (Submitted.)

Furuya, M., Wada, M. & Kadota, A. 1980 Regulation of cell growth and cell cycle by blue light in *Adiantum* gametophytes. In *The blue light syndrome* (ed. H. Senger), pp. 119–132. Berlin, Heidelberg and New York: Springer.

Goding, J. W. 1976 The chromic chloride method of coupling antigens to erythrocytes: definition of some important parameters. *J. Immunol. Meth.* **10**, 61–66.

Hahn, T.-R., Kang, S.-S. & Song, P.-S. 1980 Difference in the degree of exposure of chromophores in the Pr and Pfr forms of phytochrome. *Biochem. biophys. Res. Commun.* **97**, 1317–1323.

Hahn, T. R. & Song, P.-S. 1981 Hydrophobic properties of phytochrome as probed by 8-anilinonaphthalene-1-sulfonate fluorescence. *Biochemistry, Wash.* **20**, 2602–2609.

Hahn, T.-R. & Song, P.-S. 1982 Molecular topography of phytochrome as deduced from the tritium-exchange method. *Biochemistry, Wash.* **21**, 1394.

Haupt, W., Mörtel, G. & Winkelnkemper, I. 1969 Demonstration of different dichroic orientation of phytochrome P_R and P_{FR}. *Planta* **88**, 183–186.

Hopkins, D. W. & Butler, W. L. 1970 Immunochemical and spectroscopic evidence for protein conformational changes in phytochrome transformations. *Pl. Physiol.* **45**, 567–570.

Hunt, R. E. & Pratt, L. H. 1981 Physicochemical differences between the red- and the far-red-absorbing forms of phytochrome. *Biochemistry, Wash.* **20**, 941–945.

Inoue, Y., Konomi, K. & Furuya, M. 1982 Phototransformation of the far-red light-absorbing form of large pea phytochrome by laser flash excitation. *Pl. Cell Physiol.* **23**, 731–736.

Kadota, A., Wada, M. & Furuya, M. 1982 Phytochrome-mediated phototropism and different dichroic orientation of P_r and P_{fr} in protonemata of the fern *Adiantum capillus-veneris* L. *Photochem. Photobiol.* **35**, 533–536.

Kendrick, R. E. & Spruit, C. J. P. 1977 Phototransformations of phytochrome. *Photochem. Photobiol.* **26**, 201–214.

Kim, I.-S. & Song, P.-S. 1981 Binding of phytochrome to liposomes and protoplasts. *Biochemistry, Wash.* **20**, 5482–5489.

Köhler, G. & Milstein, C. 1975 Continuous cultures of fused cells secreting antibody of predefined specificity. *Nature, Lond.* **256**, 495–497.

Konomi, K., Furuya, M., Yamamoto, K. T., Yokota, T. & Takahashi, N. 1982 Effects of a triterpenoid saponin on spectral properties of undegraded pea phytochrome. *Pl. Physiol.* **70**, 307–310.

Laemmli, V. K. 1970 Cleavage of structural proteins during the assembly of the head of bacteriophage T4. *Nature, Lond.* **227**, 680–685.

Lagarias, J. C. & Rapoport, H. 1980 Chromopeptides from phytochrome. The structure and linkage of the P_R form of the phytochrome chromophore. *J. Am. chem. Soc.* **120**, 4821–4828.

Linschitz, H., Kasche, V., Butler, W. L. & Siegelman, H. W. 1966 The kinetics of phytochrome conversion. *J. biol. Chem.* **241**, 3395–3403.

Mason, D. W. & Williams, A. F. 1980 The kinetics of antibody binding to membrane antigens in solution and at the cell surface. *Biochem. J.* **187**, 1–20.

Mumford, F. E. & Jenner, E. L. 1966 Purification and characterization of phytochrome from oat seedlings. *Biochemistry, Wash.* **5**, 3657–3662.

Nagatani, A., Yamamoto, K. T., Furuya, M., Fukumoto, T. & Yamashita, A. 1983a Production and characterization of monoclonal antibodies of rye (*Secale cereale*) phytochrome. *Pl. Cell Physiol.* **24**. (In the press.)

Nagatani, A., Yamamoto, K. T., Furuya, M., Fukumoto, T. & Yamashita, A. 1983b Production and characterization of monoclonal antibodies of pea (*Pisum sativum*) phytochrome. (In preparation.)

Pratt, L. H. 1973 Comparative immunochemistry of phytochrome. *Pl. Physiol.* **51**, 203–209.

Pratt, L. H. 1978 Molecular properties of phytochrome. *Photochem. Photobiol.* **27**, 81–105.

Pratt, L. H. 1979 Phytochrome: function and properties. *Photochem. Photobiol. Rev.* **4**, 59–124.

Pratt, L. H. 1982a Phytochrome: the protein moiety. *A. Rev. Pl. Physiol.* **32**. (In the press.)

Pratt, L. H. 1982b Molecular properties of phytochrome and their relationship to phytochrome function. In *Strategies of plant reproduction* (ed. W. J. Meudt), pp. 117–134. Granada: Allanheld, Osmun Publishers.

Pratt, L. H. & Butler, W. L. 1970 The temperature dependence of phytochrome transformations. *Photochem. Photobiol.* **11**, 361–369.

Pratt, L. H. & Cundiff, S. C. 1975 Spectral characterization of high-molecular-weight phytochrome. *Photochem. Photobiol.* **21**, 91–97.

Pratt, L. H., Inoue, Y. & Furuya, M. 1983 Photoactivity of transient intermediates in the pathway from the red-absorbing to the far-red-absorbing form of *Avena* phytochrome as observed by a double-flash transient-spectrum analyzer. *Photochem. Photobiol.* (Submitted.)

Pratt, L. H., Shimazaki, Y., Inoue, Y. & Furuya, M. 1982 Analysis of phototransformation intermediates in the pathway from the red-absorbing to the far-red-absorbing form of *Avena* phytochrome by a multichannel transient spectrum analyzer. *Photochem. Photobiol.* **36**, 471–477.

Rice, H. V. & Briggs, W. R. 1973 Immunochemistry of phytochrome. *Pl. Physiol.* **51**, 939–945.

Rüdiger, W. 1980 Phytochrome, a light receptor of plant photomorphogenesis. *Struct. Bonding*, **40**, 101–140.

Sarkar, H. K. & Song, P.-S. 1982 Nature of phototransformation of phytochrome as probed by intrinsic tryptophan residues. *Biochemistry, Wash.* **21**, 1967–1972.

Sasaki, N., Oji, Y., Yoshizawa, T., Yamamoto, K. T. & Furuya, M. 1983a Temperature dependencies of absorption spectra of pea phytochrome and relative quantum yields of its phototransformations. *Biochim biophys. Acta* (Submitted.)

Sasaki, N., Yoshizawa, T., Yamamoto, K. T. & Furuya, M. 1983b Photochemical intermediates of pea phytochrome determined at low temperatures. *Photochem. Photobiol.* (Submitted.)

Senger, H. (ed.) 1980 *The blue light syndrome.* Berlin: Springer-Verlag.

Shimazaki, Y., Inoue, Y., Yamamoto, K. T. & Furuya, M. 1980 Phototransformation of the red-light-absorbing form of undegraded pea phytochrome by laser flash excitation. *Pl. Cell Physiol.* **21**, 1619–1625.

Shropshire, W. & Mohr, H. 1983 Photomorphogenesis. In *Encyclopedia of plant physiology (New Series)*. Berlin: Springer-Verlag. (In the press.)

Smith, H. (ed.) 1976 *Light and plant development.* London: Butterworth.

Smith, H. 1982 Light quality, photoperception, and plant strategy. *A. Rev. Pl. Physiol.* **33**, 481–518.

Smith, W. O. Jr 1981 Probing the molecular structure of phytochrome with immobilized Cibacron blue 3GA and blue dextran. *Proc. natn. Acad. Sci. U.S.A.* **78**, 2977–2980.

Song, P.-S., Sarkar, H. K., Kim, I.-S. & Poff, K. L. 1981 Primary photoprocesses of undegraded phytochrome excited with red and blue light at 77 K. *Biochim. biophys. Acta* **635**, 369–382.

Song, P.-S., Sarkar, H. K., Tabba, H. & Smith, K. M. 1982 The phototransformation of phytochrome probed by 360 MHz proton NMR spectra. *Biochem. biophys. Res. Commun.* **105**, 279–287.

Spruit, C. J. P., Kendrick, R. E. & Cooke, R. J. 1975 Phytochrome intermediates in freeze-dried tissue. *Planta* **127**, 121–132.

Stoker, B. M., McEntire, K. & Roux, S. J. 1978 Identification of tryptic chromopeptides of phytochrome on sodium dodecyl sulfate gels: implications for structure. *Photochem. Photobiol.* **27**, 597–602.

Sugimoto, T., Inoue, Y., Suzuki, H. & Furuya, M. 1983 Models for chromophore structure of phototransformation intermediates in pea phytochrome *in vitro*. *Photochem. Photobiol.* (Submitted.)

Tobin, E. M. & Briggs, W. R. 1973 Studies on the protein conformation of phytochrome. *Photochem. Photobiol.* **18**, 487–495.

Tokutomi, S., Yamamoto, K. T., Furuya, M. 1981 Photoreversible changes in hydrophobicity of undegraded pea phytochrome determined by partition in an aqueous two-phase system. *FEBS Lett.* **134**, 159–162.

Tokutomi, S., Yamamoto, K. T. & Furuya, M. 1983*a* Photoreversible proton uptake and release of pea phytochrome. *Biochim. biophys. Acta* (Submitted.)

Tokutomi, S., Yamamoto, K. T. & Furuya, M. 1983*b* Phytoreversible proton release of pea phytochrome during phototransformation of P_{659} to P_{bl}. *FEBS Lett.* (Submitted.)

Tokutomi, S., Yamamoto, K. T., Miyoshi, Y. & Furuya, M. 1982 Photoreversible changes in pH of pea phytochrome solutions. *Photochem. Photobiol.* **35**, 431–433.

Vierstra, R. D. & Quail, P. H. 1982*a* Proteolysis alters the spectral properties of 124 kdalton phytochrome from *Avena. Planta* **156**, 158–165.

Vierstra, R. D. & Quail, P. H. 1982*b* Native phytochrome: inhibition of proteolysis yields a homogeneous monomer of 124 kilodaltons from *Avena. Proc. natn. Acad. Sci. U.S.A.* **79**, 5272–5276.

Yamamoto, K. T. & Furuya, M. 1983 Spectral properties of chromophore-containing fragments prepared from pea phytochrome by limited proteolysis. *Pl. Cell Physiol.* **24**, 713–718.

Yamamoto, K. T. & Smith, W. O. Jr 1981 Alkyl and ω-amino alkyl agaroses as probes of light-induced changes in phytochrome from pea seedlings (*Pisum sativum* cv. Alaska) *Biochim. biophys. Acta* **668**, 27–34.

Discussion

W. Haupt (*Institut für Botanik und Pharmazeutische Biologie, Erlangen, F.R.G.*). In Professor Furuya's scheme with the phytochrome intermediates, I realized that only light-dependent reversions of I_{692} and I_{bl} to P_r were shown. This is in contrast to the classical Kendrick–Spruit scheme, according to which dark reversions (or dark relaxations) to P_r also occur. Is this a definite revision of the classical scheme?

M. Furuya. No. I cannot say anything on the dark reversion from our present work. What I talked about is that I_{692} and I_{bl} went back to P_r, but not to P_{fr}, by a second laser flash excitation. This is the first evidence found at room temperature, and tends to confirm the results obtained by low-temperature spectroscopy.

Phil. Trans. R. Soc. Lond. B **303**, 377–386 (1983)
Printed in Great Britain

Chemistry of the phytochrome photoconversions

By W. Rüdiger

*Botanisches Institut der Universität München, Menzinger Strasse 67,
D-8000 München 19, F.R.G.*

The photoconversions of phytochrome, $P_r \rightleftharpoons P_{fr}$, occur both *in vivo* and *in vitro*. Structural differences between P_r and P_{fr} are discussed for chromophore and apoprotein. The chemical structure of the P_r chromophore has been established. The P_{fr} chromophore was recently demonstrated to be the 15*E* isomer. The red shift of absorption to 730 nm in native P_{fr} is discussed as interaction between chromophore and apoprotein. The nature of this interaction is still unknown. Small changes in the apoprotein surface are of particular interest because they could be part of the signal chain in photoperception.

1. Introduction

An important criterion for the involvement of phytochrome in photoperception is the red–far-red photoreversibility of physiological responses. This is based on the photochromic properties of phytochrome. Phytochrome exists in two forms, P_r, absorbing at 660 nm, and P_{fr}, absorbing at 730 nm. The photoconversions can be carried out in both directions, *in vivo* and *in vitro*:

$$P_r \xrightleftharpoons[\text{730 nm}]{\text{660 nm}} P_{fr}.$$

It has long been known that phytochrome is a biliprotein (for a review on early work see Mitrakos & Shropshire (1972)). If one considers the light signal used by dark-grown plants for photoperception via P_r, it is obvious that the first step must be the absorption of photons by the bilin chromophore. This excitation leads to the modification of either the chromophore or the chromophore–apoprotein interaction, which has to be considered as the primary photoreaction. A chain of further modifications, probably dark relaxations, occur subsequently. Intermediates of this chain have been characterized by spectroscopy (Kendrick & Spruit 1977). The final product of this chain is P_{fr}, from which the biological signal starts, as generally accepted (see papers by Quail, Raven and Mohr in this symposium, but see also the paper by Smith for a critical view).

The same general scheme of primary photoreaction and subsequent dark relaxations is also true for photoconversion of the physiologically active P_{fr} to the physiologically inactive P_r. The primary photoreaction must deal with the chromophore here also, whereas dark relaxations can affect apoprotein as well as chromophore. Intermediates of the backward reaction are different from those of the forward reaction (Kendrick & Spruit 1977).

It is hoped that an understanding of the difference between P_r and P_{fr} will help to reveal the nature of the biological signal. The present paper deals with chemical properties of both phytochrome forms that are changed during the photoconversions. Emphasis is given to the chemistry of the chromophores of P_r and P_{fr}.

[31]

2. STRUCTURE AND CONFORMATION OF THE P_r CHROMOPHORE

The chemical structure of the P_r chromosphore has been elucidated in a number of independent papers (for review see Rüdiger 1980). Important steps were ultraviolet–visible spectroscopy under defined conditions (Grombein *et al.* 1975), analysis of products of oxidative degradation (Klein *et al.* 1977), identification of the cleaved chromophore with a product of total synthesis (Rüdiger *et al.* 1980), and high-resolution ¹H n.m.r. spectroscopy of a chromopeptide obtained from P_r (Lagarias & Rapoport 1980). Structure **1** has been deduced from these studies. The thioether linkage between chromophore and peptide chain was also revealed by the above studies

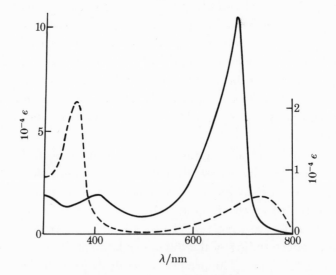

1

Bile pigments are flexible molecules, which can exist in a great number of conformations. Theoretical calculations (Chae & Song 1975; Sugimoto *et al.* 1976; Pasternak & Wagniére 1979) predicted a dramatic influence of the conformation upon spectral properties especially on the oscillator strength ratio of red to blue absorption bands. Examples of the experimental verification of these predictions are given in figure 1. The examples are bile pigments in which the conformations are fixed by chemical 'cross-linking'. The fixed conformation is closed (or

FIGURE 1. Electronic spectra of bile pigment cations with chemically fixed conformations. ---, Spectrum of **2** in acidic CHCl₃ (right ordinate) (redrawn from Falk & Thirring 1981); ——, spectrum of **3** in acidic methanol (left ordinate) (Scheer & Kufer, unpublished results).

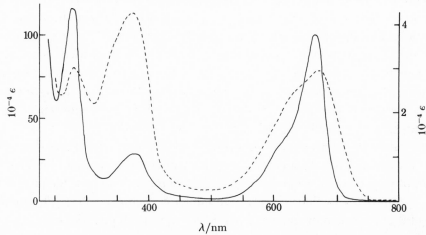

helical) in **2** and open (or extended) in **3**. In accordance with the prediction, **2** has a small and **3** a large oscillator strength ratio of red (700–750) nm to blue (350–400 nm) bands.

A comparison of the absorption spectra of the P_r chromopeptide and native P_r (figure 2) demonstrates that the chromophore has a more closed conformation in the former and a more extended conformation in the latter. It can be assumed that the (native) protein fixes a conformation of the chromophore that is otherwise unfavourable. A semicircular (or semi-extended) conformation has been calculated for native P_r (Song *et al.* 1979). Denaturation of P_r (e.g. with 8 M urea or 6 M guanidinium chloride) leads to the same spectral properties as for the P_r chromopeptide (Grombein *et al.* 1975). This resembles the situation in phycocyanin, an accessory biliprotein pigment of photosynthesis in blue-green and red algae (Scheer & Kufer 1977). A difference concerns the stability: denatured phycocyanin can be renatured, whereas renaturation is not possible with denatured P_r.

FIGURE 2. Electronic spectra of large phytochrome in the P_r form (——, left ordinate) and P_r chromopeptide (---, right ordinate). Phytochrome from oat ($A_{667}/A_{280} = 0.87$) with $\epsilon_{667\,nm} = 10.2 \times 10^4$ (after Roux *et al.* 1982). Independent measurements in our laboratory yielded $\epsilon_{665\,nm} = 10.9$–$11.8 \times 10^4$ per phytochromobilin chromophore (Brandlmeier *et al.* 1981). P_r chromopeptide in 1 % aqueous formic acid (after Thümmler & Rüdiger 1983).

It can be speculated that plants developed strong red bands in the chromophores of phycocyanin and phytochrome for specific use of red light and minimal interference of blue–u.v. light. However, the functions seem to be quite different. Phycocyanin has been optimized for energy transfer. The protein stabilizes the chromophore, i.e. it inhibits photochemical and other radiationless deactivation processes. Absorbed light is emitted as fluorescence with high quantum efficiency. Phytochrome has been optimized for photoconversion. The absolute quantum yield for both $P_r \rightarrow P_{fr}$ and $P_{fr} \rightarrow P_r$ transformations has been determined as 0.17 (Pratt 1975). This high quantum efficiency implies that other radiationless deactivation processes are inhibited. Phytochrome exhibits an almost complete lack of fluorescence at room temperature (Song et al. 1979).

3. Structure of the P_{fr} chromophore

Early investigations demonstrated that the P_{fr} chromophore is less stable than the P_r chromophore; the native protein serves to stabilize the chromophore. The relevant literature contains much speculation on possible structures of the P_{fr} chromophore (recently summarized by Lagarias & Rapoport 1980). Some of these speculations could be ruled out by rather simple experiments. Oxidative degradation revealed that the β side chain of the chromophore and its linkage to the protein are identical in P_{fr} and P_r (Klein et al. 1977). This ruled out changes of double bonds in the side chains (Siegelman et al. 1968) or the cleavage of the thioether linkage (Song et al. 1979) during the $P_r \rightarrow P_{fr}$ transformation. Early spectroscopic measurements (Grombein et al. 1975) demonstrated differences between denatured P_{fr} and denatured P_r; the meaning of these results were questioned because of possible artefacts during denaturation (Kendrick & Spruit 1977). A difference between the P_{fr} chromophore and the P_r chromophore could finally be established by preparation of chromopeptides from both phytochrome forms.

(a) Preparation and isolation of chromopeptides from P_{fr}

A small chromopeptide was at first isolated by Fry & Mumford (1971) after pepsin digestion of phytochrome. Those authors did not find any difference in the composition of chromopeptide starting with either P_r or P_{fr}. Such a difference between P_{fr} chromopeptides and P_r chromopeptides was, however, found if pepsin digestion was carried out under carefully controlled conditions (Thümmler et al. 1981). The conditions had to be obeyed because otherwise the P_{fr} peptide was easily transformed into the P_r peptide. Digestion and isolation had to be performed in darkness or under green safety-light and between pH 2 and 4; the incubation time at 37 °C had to be kept at a minimum. It was shown that white light and strong acids and bases catalyse the transformation into the P_r peptide. A slow, temperature-dependent 'reversion' of the P_{fr} peptide occurred also in the dark. Isolation included column chromatography on Biogel P-10 and silica gel with step gradients of acids (acetic acid, formic acid or, later, diluted hydrochloric acid) (Thümmler & Rüdiger 1983). This method allowed separation of P_{fr} peptides from colourless peptides and from P_r peptides but not the separation of single P_{fr} chromopeptides from each other. This heterogeneity was confirmed by amino acid analysis (Thümmler & Rüdiger 1983).

(b) Chemical and spectroscopic properties: comparison with model chromophores

The ultraviolet–visible spectra of the P_{fr} chromopeptide and the P_r chromopeptide obtained from it by photoconversion are shown in figure 3. The spectra resemble those of Z, E isomeric biliverdin chromophores (Thümmler et al. 1981). The hypsochromic shift of the red band in the E configurated model compounds was explained by partial uncoupling of one pyrrole ring from the conjugated system due to steric hindrance (Falk et al. 1978). E configurated model

[34]

compounds are unstable, like the P_{fr} chromopeptide. Irradiation of the cation with white light, and contact with strong acids and bases yield the Z configurated isomers.

The best model for the P_{fr} chromopeptide proved to be a chromopeptide from phycocyanin (Thümmler & Rüdiger 1983). Pepsin digestion of phycocyanin yielded chromopeptides which contain the all-Z configurated chromophore **4**. This chromophore can be converted to its E configurated isomer by a procedure outlined in figure 4. The decision between the two possible

FIGURE 3. Electronic spectra of P_{fr} chromopeptide (——) and P_r chromopeptide (– – –) in 1 % aqueous formic acid (after Thümmler & Rüdiger 1983).

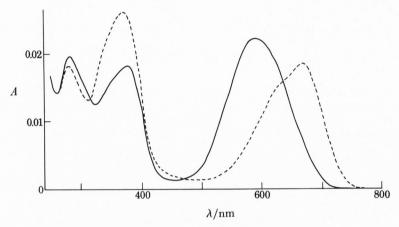

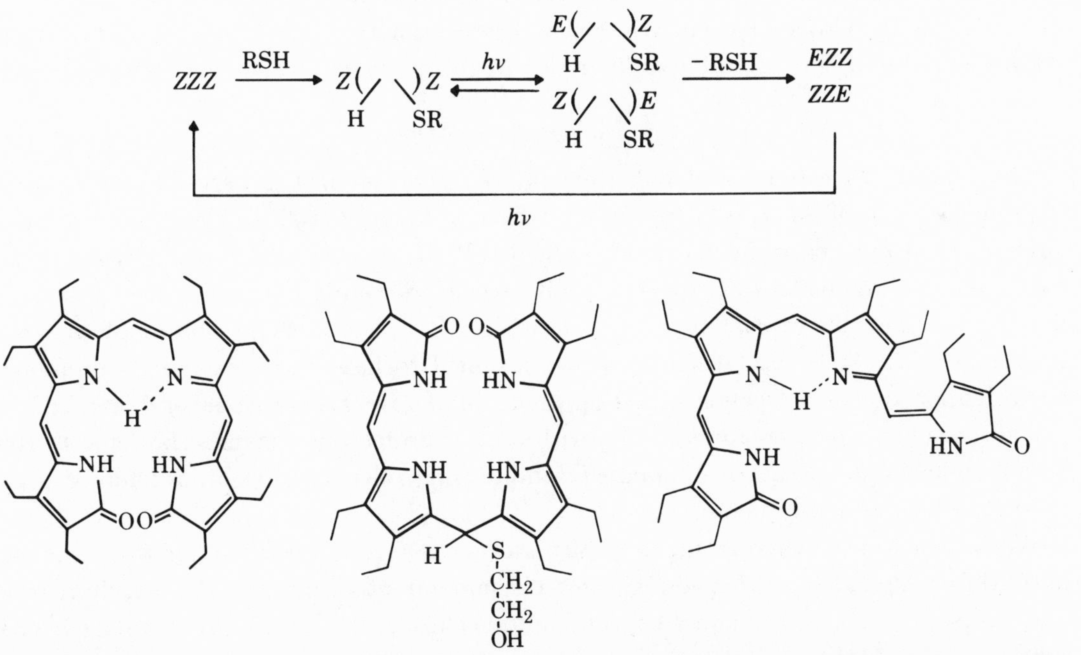

FIGURE 4. Reaction sequence for preparation of E configurated bile pigments by a procedure of Falk *et al.* (1980). Structures are given for octaethylbiliverdin (from left to right): all-Z educt, adduct, E configurated product.

structures, the $4E$ and the $15E$ configurated chromophore, was obtained by high-resolution ¹H n.m.r. spectroscopy: only the $15E$ configurated phycocyanin chromopeptide **5** is formed by the procedure of figure 4 (Thümmler *et al.* 1983). The methine signals are particularly

[35]

valuable for localization of the isomerized methine double bond (Rüdiger *et al.* 1983). The high resolution ^{1}H n.m.r. spectrum of the P_{fr} chromopeptide revealed that this also contains the $15E$ configuration (Rüdiger *et al.* 1983; Thümmler *et al.* 1983). Because **6** is the structure of the P_{fr} chromophore in the chromopeptide it can be concluded that the Z, E (or *cis, trans*) isomerization at C15 is an essential step of the $P_r \rightleftharpoons P_{fr}$ photoconversion.

6 R=CH=CH$_2$
5 R=C$_2$H$_5$

4

It should be mentioned here that **6** is more reactive towards oxidation and reduction than the Z configurated isomer **1**. Therefore oxidation or reduction of **6** yields some **1** besides the proper oxidation or reduction products (Thümmler *et al.* 1981). The catalysis of dark reversion of native P_{fr} by reductants (Mumford & Jenner 1971) has been explained by the same mechanism (Thümmler *et al.* 1981). The chromophore is more easily oxidized in native P_{fr} than in native P_r. This has been interpreted as difference in the accessibility of the chromophore (Hahn *et al.* 1980) but the above difference in reactivity has yet to be considered in this connection.

(c) *The role of the apoprotein*

The typical absorption band of native P_{fr} ($\lambda_{max} = 730$ nm) is only formed at the last intermediate step of the $P_r \rightarrow P_{fr}$ photoconversion (Pratt *et al.* 1982). The direct precursor is a relatively weakly absorbing intermediate (called P_{bl} or *meta*-Rb) that has a broad absorption maximum at about 650 nm (Cross *et al.* 1968). A 'closed' conformation has been proposed for this intermediate (Burke *et al.* 1972), which implies that it contains less chromophore–protein interaction than P_r or P_{fr}, which both are assumed to have more extended chromophore conformations due to interaction with the protein. Similar weakly absorbing or 'bleached' forms can also be obtained by treatment of P_{fr} with urea or proteolytic enzymes (Butler *et al.* 1964), with 8-anilinonaphthalene-1-sulphonate (Hahn & Song 1981) or by rigorous removal of water from P_{fr} (Balangé 1974; Tobin *et al.* 1973).

We have checked whether the intermediate *meta*-Rb and bleached phytochrome forms derived from P_{fr} contain the E configurated chromophore **6** or whether the bleaching process is accompanied by a reversion to the Z configurated chromophore **1**. For this purpose, bleached solutions were acidified. The solution then contained the chromophore cation linked to denatured protein. Starting with P_{fr}, this treatment yielded a product spectroscopically identical with the P_{fr} chromopeptide. The experiment is illustrated here with *meta*-Rb (figure 5). It turned out that all bleached phytochrome preparations still contained the chromophore **6**, which could be transformed into **1** by irradiation of the acidic solution with white light. Apparently, the $Z \rightarrow E$ isomerization is not related directly with the absorption shift to

730 nm. It must therefore be one of the early intermediate steps of the $P_r \rightarrow P_{fr}$ transformation. By analogy with rhodopsin (Wald 1968), it could be a good candidate for the primary photoreaction. However, the deuterium isotope effect found for the $P_r \rightarrow P_{fr}$ phototransformation has been interpreted in terms of the primary reaction involving an intramolecular proton transfer (Sarkar & Song 1981). The exact chemistry of the primary photoreaction remains to be elucidated.

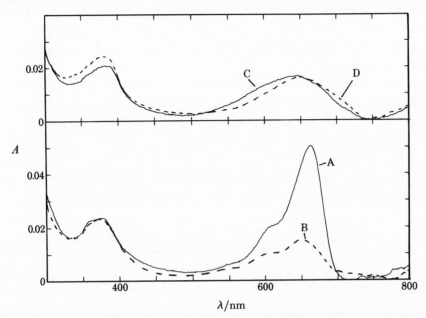

FIGURE 5. Electronic spectra of phytochrome preparations in 66 vol. % aqueous glycerol at $-40\,°C$. Lower traces: A, phytochrome in the P_r form, B, the same after saturating irradiation at 660 nm. Upper traces: C, B after acidification with HCl (this corresponds to the P_{fr} chromopeptide); D, C after irradiation with white light (this corresponds to the P_r chromopeptide) (Thümmler & Rüdiger, unpublished).

A consequence of the $Z \rightarrow E$ isomerization is the prediction that chromophores 1 and 6 should have different positions in relation to the bulk apoprotein. Different dichroic orientation of P_r and P_{fr} in cells of *Mougeotia* and other organisms has been known since a long time (Haupt 1972). It was not known, however, whether this signifies different orientation of the whole phytochrome molecule in the membrane or of the chromophore within the apoprotein. The latter possibility was more recently supported by chemical and spectroscopical investigations on phytochrome in solution (Hahn *et al.* 1980; Hahn & Song 1981) and by investigation of immobilized phytochrome with polarized light (Sundquist & Björn 1983).

4. DIFFERENCES BETWEEN P_r AND P_{fr}: THE APOPROTEIN

The photoconversion of phytochrome could not lead to photoperception by plants if photochemical changes were restricted to the chromophore and eventually to the immediate protein environment of the chromophore. Differences between P_r and P_{fr} other than spectral shifts were postulated early in phytochrome research, but many attempts to establish such differences failed, as critically reviewed by Pratt (1978, 1982). In summary, no drastic conformational changes of the protein were found. The biological significance of minor changes (probably at the protein surface) remains unclear because only proteolytically degraded phytochrome was studied in early investigations. The isolation of native phytochrome (124 kDa) has only recently been described (Vierstra & Quail 1982; Kerscher & Nowitzki 1982). Some investigations with partly degraded phytochrome (114–118 kDa) will be considered here because these results are probably also relevant for native phytochrome.

Hunt & Pratt (1981) established differences in the accessibility of several amino acids that point to a conformational change (eventually small) at the surface. This may be related to the finding of a hydrophobic surface area in P_{fr} that is not found in P_r (Hahn & Song 1981; Tokutomi et al. 1981). Photoreversible pH changes caused by unbuffered phytochrome solutions (Tokutomi et al. 1982) could be due to the exposure of charged amino acid residues by the photoconversion.

All of these changes could occur in only a restricted area of the protein surface because no difference between P_r and P_{fr} could be detected with immunological methods (see Pratt 1978). The changes are currently of particular interest: they could be part of the signal chain leading from P_{fr} to the biological responses in photoperception.

The cited work from the laboratory of the author was supported by the Deutsche Forschungsgemeinschaft, Bonn.

REFERENCES

Balangé, A. P. 1974 Spectral changes of phytochrome in glycerol solution. Physiol. vég. 12, 95–105.

Brandlmeier, T., Scheer, H. & Rüdiger, W. 1981 Chromophore content and molar absorptivity of phytochrome in the P_r form. Z. Naturf. C 36, 431–439.

Burke, M. J., Pratt, D. C. & Moscowitz, A. 1972 Low-temperature absorption and circular dichroism studies of phytochrome. Biochemistry, Wash. 11, 4025–4031.

Butler, W. L., Siegelmann, H. W. & Miller, C. O. 1964 Denaturation of phytochrome. Biochemistry, Wash. 3, 851–857.

Chae, Q. & Song, P. S. 1975 Linear dichroic spectra and fluorescence polarization of biliverdin. J. Am. chem. Soc. 97, 4176–4179.

Cross, D. R., Linschitz, H., Kasche, V. & Tenenbaum, J. 1968 Low-temperature studies on phytochrome: light and dark reactions in the red to far-red transformation and new intermediate forms of phytochrome. Proc. natn. Acad. Sci. U.S.A. 61, 1095–1101.

Falk, H., Grubmayr, K., Haslinger, E., Schlederer, T. & Thirring, K. 1978 Beiträge zur Chemie der Pyrrolpigmente. 25. Mitt. Die diastereomeren (geometrisch isomeren) Biliverdindimethylester-Struktur, Konfiguration und Konformation. Mh. Chem. 109, 1451–1473.

Falk, H., Müller, N. & Schlederer, T. 1980 Beiträge zur Chemie der Pyrrolpigmente. 35. Mitt. Eine regioselektive, reversible Addition an Bilatriene-abc. Mh. Chem. 111, 159–175.

Falk, H. & Thirring, K. 1981 Beiträge zur Chemie der Pyrrolpigmente. XXXVII. Überbrückte Gallenpigmente: N_{21}-N_{24}-Methylenactivbiliverdin-IV_8 und N_{21}-N_{24}-Methylenactivbilirubin-IV_8. Tetrahedron 37, 761–766.

Fry, K. T. & Mumford, F. E. 1971 Isolation and partial characterization of a chromophore–peptide fragment from pepsin digest of phytochrome. Biochem. biophys. Res. Commun. 45, 1466–1473.

Grombein, S., Rüdiger, W. & Zimmermann, H. 1975. The structures of the phytochrome chromophore in both photoreversible forms. Hoppe-Seyler's Z. physiol. Chem. 356, 1709–1714.

Hahn, T.-R., Kang, S.-S. & Song, P.-S. 1980 Difference in the degree of exposure of chromophores in the P_r and P_{fr} forms of phytochrome. Biochem. biophys. Res. Commun. 97, 1317–1323.

Hahn, T.-R. & Song, P.-S. 1981 Hydrophobic properties of phytochrome as probed by 8-anilinonaphthalene-1-sulfonate fluorescence. Biochemistry, Wash. 20, 2602–2609.

Haupt, W. 1972 Localization of phytochrome within the cell. In Phytochrome (ed. K. Mitrakos & W. Shropshire, Jr.), pp. 553–569. London and New York: Academic Press.

Hunt, R. E. & Pratt, L. H. 1981 Physiochemical differences between the red- and the far-red-absorbing forms of phytochrome. Biochemistry, Wash. 20, 941–945.

Kendrick, R. E. & Spruit, C. J. P. 1977 Phototransformations of phytochrome. Photochem. Photobiol. 26, 201–214.

Kerscher, L. & Nowitzki, S. 1982 Western blot analysis of a lytic process in vitro specific for the red light absorbing form of phytochrome. FEBS Lett. 146, 173–176.

Klein, G., Grombein, S. & Rüdiger, W. 1977 On the linkage between chromophore and protein in biliproteins. VI. Structure and protein linkage of the phytochrome chromophore. Hoppe-Seyler's Z. physiol. Chem. 358, 1077–1079.

Lagarias, J. C. & Rapoport, H. 1980 Chromopeptides from phytochrome. The structure and linkage of the P_r form of the phytochrome chromophore. *J. Am. chem. Soc.* **102**, 4821–4828.

Mitrakos, K. & Shropshire, W. Jr (eds) 1972 *Phytochrome.* London & New York: Academic Press.

Mumford, F. E. & Jenner, E. L. 1971 Catalysis of the phytochrome dark reaction by reducing agents. *Biochemistry, Wash.* **10**, 98–101.

Pasternak, R. & Wagniére, G. 1979 Possible interpretation of long-wavelength spectral shifts in phytochrome P_r and P_{fr} forms. *J. Am. chem. Soc.* **101**, 1662–1667.

Pratt, L. H. 1975 Photochemistry of high molecular weight phytochrome in vitro. *Photochem. Photobiol.* **22**, 33–36.

Pratt, L. H. 1978 Molecular properties of phytochrome. *Photochem. Photobiol.* **27**, 81–105.

Pratt, L. H. 1982 Phytochrome: the protein moiety. *A. Rev. Pl. Physiol.* **33**, 557–582.

Pratt, L. H., Shimazaki, Y., Inone, Y. & Furuya, M. 1982 Analysis of phototransformation intermediates in the pathway from the red-absorbing to the far-red-absorbing form of *Avena* phytochrome by multichannel transient spectrum analyzer. *Photochem. Photobiol.* **36**, 471–477.

Roux, S. J., McEntire, K. & Brown, W. E. 1982 Determination of extinction coefficients of oat phytochrome by quantitative amino acid analyses. *Photochem. Photobiol.* **35**, 537–544.

Rüdiger, W. 1980 Phytochrome, a light receptor of plant photomorphogenesis. *Struct. Bonding* **40**, 101–140.

Rüdiger, W., Brandlmeier, T., Blos, I., Gossauer, A. & Weller, J.-P. 1980 Isolation of the phytochrome chromophore. The cleavage reaction with hydrogen bromide. *Z. Naturf.* **35c**, 763–769.

Rüdiger, W., Thümmler, F., Cmiel, E. & Schneider, S. 1983 Chromophore structure of the physiologically active form (P_{fr}) of phytochrome. *Proc. natn. Acad. Sci. U.S.A.* (In the press.)

Sarkar, H. K. & Song, P.-S. 1981 Phototransformation and dark reversion of phytochrome in D_2O. *Biochemistry, Wash.* **20**, 4315–4320.

Scheer, H. & Kufer, W. 1977 Conformational studies on C-phycocyanin from *Spirulina platensis*. *Z. Naturf.* **32c**, 513–519.

Siegelman, H. W., Chapman, D. J. & Cole, W. J. 1968 The bile pigments of plants. In *Porphyrins and related compounds* (ed. T. W. Goodwin), pp. 107–120. London and New York: Academic Press.

Song, P.-S., Chae, Q. & Gardner, J. G. 1979 Spectroscopic properties and chromophore conformations of the photomorphogenic receptor: phytochrome. *Biochim. biophys. Acta* **576**, 479–495.

Sugimoto, T., Ishikawa, K. & Suzuki, H. 1976 On the models for phytochrome chromophore. III. *J. phys. Soc. Japan* **40**, 258–266.

Sundquist, C. & Björn, L. O. 1983 Light-induced linear dichroism in photoreversibly photochromic sensor pigments. II. Chromophore rotation in immobilized phytochrome. *Photochem. Photobiol.* **37**, 63–75.

Thümmler, F., Brandlmeier, T. & Rüdiger, W. 1981 Preparation and properties of chromopeptides from the P_{fr} form of phytochrome. *Z. Naturf.* **36c**, 440–449.

Thümmler, F. & Rüdiger, W. 1983 Models for photoreversibility of phytochrome: Z, E isomerization of chromopeptides from phycocyanin and phytochrome. *Tetrahedron.* (In the press.)

Thümmler, F., Rüdiger, W., Cmiel, E. & Schneider, S. 1983 Chromopeptides from phytochrome and phycocyanin. NMR studies of the P_{fr} and P_r chromophore of phytochrome and E, Z isomeric chromophores of phycocyanin. *Z. Naturf.* **38c**, 359–368.

Tobin, E. M., Briggs, W. R. & Brown, P. K. 1973 The role of hydration in the phototransformation of phytochrome. *Photochem. Photobiol.* **18**, 497–503.

Tokutomi, S., Yamamoto, K. T. & Furuya, M. 1981 Photoreversible changes in hydrophobicity of undegraded pea phytochrome determined by partition in an aqueous two-phase system. *FEBS Lett.* **134**, 159–162.

Tokutomi, S., Yamamoto, K. T., Miyoshi, Y. & Furuya, M. 1982 Photoreversible changes in pH of pea phytochrome solutions. *Photochem. Photobiol.* **35**, 431–433.

Vierstra, R. D. & Quail, P. H. 1982 Native phytochrome: inhibition of proteolysis yields a homogeneous monomer of 124 kilodaltons from *Avena*. *Proc. natn. Acad. Sci. U.S.A.* **79**, 5272–5276.

Wald, G. 1968 The molecular basis of visual excitation. *Nature, Lond.* **219**, 800–807.

Discussion

R. BONNETT (*Department of Chemistry, Queen Mary College, London, U.K.*). With respect to the question of whether the protein or the linear tetrapyrrole undergoes a photoinduced change, it should be borne in mind that although the protein can undoubtedly perturb the chromophore, it is the linear tetrapyrrole that absorbs the photon. In other words, of all the chemical groups present, it is the linear tetrapyrrole that becomes excited first. Of course, it may pass on its excitation energy, but the geometrical isomerization of the excited linear tetrapyrrole is a very plausible route, and is strongly supported by Professor Rüdiger's present evidence.

My questions have to do with the chromopeptides. Firstly, what is the nature of the chromopeptide: how may amino acids are present? Secondly, do I understand correctly from the nuclear magnetic spectra at 500 MHz that the sample of P_{fr} chromopeptide subjected to the irradiation experiment is an approximately 1:1 mixture of P_r and P_{fr} chromopeptides?

W. RÜDIGER. The chromopeptides from P_{fr} were prepared by the same method as previously from P_r (Fry & Mumford 1971; Lagarias & Rapoport 1980). The main product in that case was an undecapeptide containing some smaller peptides. Our amino acid analysis (Thümmler *et al.* 1983) confirmed that the same peptides were obtained from P_{fr}. Further fractionation of P_{fr} chromopeptides was unfeasible owing to rapid dark reversion to P_r chromopeptides. This is also the reason why the sample of P_{fr} chromopeptide contained considerable amounts of P_r chromopeptide at the time of the n.m.r. measurement. The exact percentage cannot easily be determined but, according to electronic and n.m.r. spectra, the sample contained somewhat more P_{fr} than P_r chromopeptide.

Phil. Trans. R. Soc. Lond. B **303**, 387–402 (1983)
Printed in Great Britain

Phytochrome: molecular properties and biogenesis

By P. H. Quail, J. T. Colbert, H. P. Hershey and R. D. Vierstra
Department of Botany, University of Wisconsin, 139 *Birge Hall, Madison, Wisconsin* 53706, *U.S.A.*

Native *Avena* phytochrome, recently shown to have a monomeric molecular mass of 124 kDa, has molecular properties that differ significantly from those of the extensively characterized '120' kDa or 'large' phytochrome preparations now known to contain a mixture of proteolytically degraded 118 and 114 kDa polypeptides. For example, 124 kDa phytochrome has a blocked N-terminus, a P_{fr} λ_{max} of 730 nm, a higher photostationary state in red light (86 % P_{fr}), exhibits no dark reversion and shows no differential reactivity of P_r and P_{fr} toward a chemical probe of hydrophobic domains. The data indicate that the proteolytically removed 6–10 kDa polypeptide segment(s) is critical to the spectral and structural integrity of the photoreceptor; that at least part of the cleaved domain is located at the N-terminus of the molecule; that this domain influences the chemical reactivity of the chromophore with the external medium; and that a current hypothesis that P_r–P_{fr} photoconversion results in the exposure of a hydrophobic domain on the molecule is inconsistent with the properties of native phytochrome.

Phytochrome has been found to exert rapid negative feedback control over the level of its own translatable mRNA. P_{fr} formation in etiolated tissue causes a decline in translatable phytochrome mRNA that is detectable within 15–30 min and that results in more than a 95 % reduction within 2 h. Less than 1 % P_{fr} is sufficient to induce 60 % of the maximum response, which is saturated at 20 % P_{fr} or less. The rapidity of this autoregulatory control makes phytochrome itself an attractive system for investigating phytochrome-regulated gene expression.

A project to clone phytochrome complementary DNA (cDNA) has been initiated. A major obstacle in this work has been the unexpectedly low abundance of phytochrome mRNA, less than 0.005 % of the poly(A) RNA in etiolated tissue. cDNA made from poly(A) RNA enriched *ca.* 200-fold in phytochrome mRNA has been cloned and bacterial colonies have been screened with a synthetic oligodeoxynucleotide hybridization probe. The sequence of this probe was derived from a known partial amino acid sequence of the phytochrome protein. Difficulties encountered with this approach are discussed.

Introduction

We have been investigating the molecular properties and biogenesis of phytochrome for the ultimate purpose of understanding the mechanism by which the photoreceptor regulates plant development. This paper describes three facets of the investigation.

The phytochrome molecule

Efforts to elucidate the mechanism of phytochrome action have focused considerable attention on the physicochemical properties of the chromoprotein and on the molecular differences between P_r and P_{fr} (see Pratt 1982). The success of this approach clearly depends on the isolation of phytochrome in an undegraded, undenatured form. Initial attempts in the mid-1960s to purify the photoreceptor yielded a photoreversible chromoprotein with an apparent molecular mass of 60 kDa (Mumford & Jenner 1966), and a considerable amount of

[41]

information on its properties was assembled (Briggs & Rice 1972; Pratt 1979, 1982). In the early 1970s, however, Briggs and coworkers isolated a molecule with a monomer size of about 120 kDa and demonstrated that the 60 kDa species was derived proteolytically from the larger polypeptide during purification (Gardner *et al.* 1971; Rice & Briggs 1973; Rice *et al.* 1973). In the subsequent period it was widely accepted that the *ca.* 120 kDa monomer ('large' phytochrome) represented the undegraded, native chromoprotein, and once again much information on its molecular properties was accumulated (Pratt 1979, 1982).

Recently, however, we have shown that phytochrome from etiolated *Avena* is a homogeneous species with a monomeric molecular mass of 124 kDa, some 6–10 kDa larger than the heterogeneous mix of 118 and 114 kDa polypeptides that actually compose the '120' kDa preparations (Quail *et al.* 1981; Vierstra & Quail 1982a). Time-course and inhibitor studies have provided strong evidence that the 118 and 114 kDa species are derived from the 124 kDa molecule by post-homogenization proteolysis and that the P_r form is much more susceptible to this proteolysis than the P_{fr} form. Similar limited proteolysis has now also been documented in extracts of rye, corn, pea and zucchini seedlings, indicating the generality of the problem (Vierstra & Quail, unpublished; Kerscher & Nowitzki 1982). Because previous purification protocols have invariably stipulated rigorous maintenance of phytochrome in the P_r form throughout the procedure, it is highly likely that most of the data collected on the molecular properties of the purified photoreceptor have been obtained with partly degraded preparations (Pratt 1979, 1982).

Three lines of evidence indicate that 124 kDa phytochrome represents the native monomer in *Avena* and is not yet another proteolytic degradation product of a still larger polypeptide. First, the spectral properties, including the peak absorbance position (λ_{max}) of P_{fr}, are the same as those determined for phytochrome *in vivo* by difference spectroscopy (Vierstra & Quail 1982b). Second, like *Avena* phytochrome *in vivo*, the 124 kDa molecule exhibits no dark reversion (Vierstra & Quail 1983a). Third, the mobility of 124 kDa phytochrome is indistinguishable from that of the *in vitro* translation product of phytochrome mRNA, indicating the absence of both *in vivo* proteolytic processing and post-homogenization proteolysis (Bolton & Quail 1982). Thus the properties determined for the 124 kDa molecule would appear likely to reflect those of undegraded, undenatured phytochrome.

While at first sight the reduction in molecular mass occurring upon proteolytic conversion of the 124 kDa to the 118/114 kDa species might appear small and potentially unimportant, the evidence so far available indicates that a large number of phytochrome properties are altered significantly by this degradation (summarized in table 1). As well as changing the pI of the molecule (Vierstra & Quail 1982a), progressive proteolysis *in vitro* has been shown, by difference spectroscopy in crude extracts, to lead to a shift in the λ_{max} of P_{fr} from 730 to 722 nm (figure 1) (Vierstra & Quail 1982b). This observation explains the previously longstanding discrepancy between the P_{fr} λ_{max} measured *in vivo* and that of purified phytochrome (Everett & Briggs 1970). The data also provide a molecular interpretation for the empirical observations of Epel and coworkers, who found that phytochrome extracted as P_{fr} had a P_{fr} λ_{max} similar to that *in vivo* whereas extraction or incubation, or both, as P_r caused a shift in P_{fr} λ_{max} by about 10 nm to shorter wavelengths (Epel 1981; Baron & Epel 1982). These workers had interpreted their data to represent some form of 'activation' and 'deactivation' in the P_{fr} and P_r forms respectively.

Studies with purified 124 kDa phytochrome, obtained with a recently developed procedure

(Vierstra & Quail 1983 a), indicate additional spectral differences between the native and partly degraded species. The absorbance spectra of purified 124 kDa phytochrome (figure 2) differ from those of purified 118/114 kDa preparations in the P_{fr} λ_{max} positions both in the far red and in the blue regions and exhibit enhanced absorbance of P_{fr} at 730 nm relative both to the 673 nm shoulder of the red-generated spectrum and to the maximum absorbance of P_r at 666 nm (table 1). Purified 124 kDa phytochrome also exhibits negligible dark reversion with or without dithionite and has an enhanced quantum yield for the $P_r \rightarrow P_{fr}$ photoconversion relative to 118/114 kDa phytochrome and a photoequilibrium in red light of 86 % P_{fr}, which is substantially higher than prevous reports for *Avena* (Vierstra & Quail 1983 a, b; Pratt 1975). Differences in the circular dichroism (c.d.) spectra, indicative of differences in molecular structure between the 124 kDa and 118/114 kDa species, have also been detected (Vierstra, Hahn, Sarkar, Song & Quail, unpublished).

TABLE 1. PREVIOUS AND REVISED PHYTOCHROME PROPERTIES

	previous	revised	references
monomer molecular mass/kDa	118/114	124	1
pI	5.8/6.0	5.9	1
N-terminus	Lys, Ala	blocked	2,3
P_r λ_{max} (red)/nm	667	666	3, 4, 5, 6
P_{fr} λ_{max} (far red)/nm	722–724	730	3, 4, 5, 6
P_r λ_{max} (blue)/nm	382	379	3, 4
P_{fr} λ_{max} (blue)/nm	390	400	3, 4
$A^{P_{fr}}_{\lambda_{max}}/A^{P_{fr}}_{shoulder}$	0.92–1.13	1.45	3, 4, 6
$A^{P_{fr}}_{\lambda_{max}}/A^{P_r}_{\lambda_{max}}$	0.43–0.48	0.58	3, 4, 6
spectral change ratio, $\Delta A_{red}/\Delta A_{far\,red}$	1.24–1.35	1.07	5, 7
P_r quantum yield (ϕ_r)	0.11	0.17	6, 8
P_{fr} quantum yield (ϕ_{fr})	0.12	0.10	6, 8
photoequilibrium in red light	0.75–0.79	0.86	6, 8
dark reversion *in vitro*	yes	minimal	3, 9
effect of ANS on photoconversion rate	altered	no effect	10, 11
Trp phosphorescence-quench by chromophore	$P_r > P_{fr}$	$P_r = P_{fr}$	11, 12
tetranitromethane bleaching $P_{fr} > P_r$	50-fold	8-fold	13, 14
c.d. spectra ΔA, 190–220 nm	$P_r = P_{fr}$	$P_r < P_{fr}$	11, 15
$\quad\quad\Delta A$ (rel. units), 700–750 nm	2	1	11, 16

References: 1, Vierstra & Quail (1982a); 2, Hunt & Pratt (1980a); 3, Vierstra & Quail (1983a); 4, Hunt & Pratt (1979b); 5, Vierstra & Quail (1982b); 6, Vierstra & Quail (1983b); 7, Pratt & Cundiff (1975); 8, Pratt (1975); 9, Pike & Briggs (1972); 10, Hahn & Song (1981); 11, Vierstra, Hahn, Sarkar, Song & Quail (unpublished); 12, Sarkar & Song (1982); 13, Hahn *et al.* (1980); 14, Hahn *et al.* (1983); 15, Tobin & Briggs (1973); 16, Song *et al.* (1979).

Experiments designed to probe the molecule for differences in the P_r and P_{fr} forms provide further evidence of molecular changes resulting from the limited proteolysis. Neither the effect of 8-anilinonaphthalene-1-sulphonate (ANS) on photoconversion rate (Hahn & Song 1981) nor the enhanced quenching of tryptophan phosphorescence by the chromophore in the P_r form (Sarkar & Song 1982) previously reported for the 118/114 kDa species is observed with 124 kDa phytochrome (Vierstra, Hahn, Sarkar, Song & Quail, unpublished). These data call into question the hypothesis of Song and coworkers (Hahn & Song 1981; Sarkar & Song 1982) that P_r–P_{fr} photoconversion results in movement of the chromophore relative to the protein with the exposure of a hydrophobic domain on the molecule. The rate of bleaching by tetranitromethane is 50-fold higher for P_{fr} than for P_r with 118/114 kDa phytochrome but only

8-fold higher with the 124 kDa molecule (Hahn *et al.* 1983). This result indicates that the chromophore is less reactive with the surrounding medium in the P_{fr} form of 124 kDa phytochrome than of 118/114 kDa phytochrome. The data do not permit a distinction between whether this difference represents a change in the intrinsic chemical reactivity of the chromophore or a change in spatial location or both.

The amino acid composition of 124 kDa phytochrome differs only minimally in molar proportions from that of 118/114 kDa preparations (Vierstra & Quail 1983 a). The minor apparent differences that have been measured indicate a possible tendency toward slight depletion of Tyr, Phe, Val and Asx and slight enrichment of Thr, Arg and Met in the

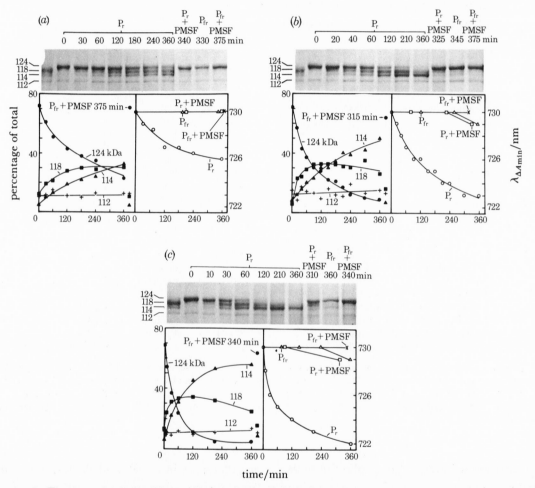

FIGURE 1. Time-course of the change in electrophoretic mobility and P_{fr} λ_{max} during incubation of *Avena* phytochrome as P_r in crude extracts from etiolated seedlings. Phytochrome was extracted in the P_{fr} form and then either retained as P_{fr} or reconverted to P_r in the 48000 g supernatant from the extract before incubation at 2 °C (*a*), 10 °C (*b*) and 20 °C (*c*) with or without the addition of 2 mM PMSF. At the times indicated difference spectra ($P_r - P_{fr}$) were recorded and phytochrome was immunoprecipitated in the P_{fr} form (in all cases) with antiphytochrome IgG-coated *S. aureus* cells (Ivarie & Jones 1979). The immunoprecipitates were subjected to sodium dodecyl sulphate polyacrylamide gel electrophoresis in 5% (by mass) acrylamide gels (Laemmli 1970), and the percentage of phytochrome in each molecular mass species (124 (●), 118 (■), 114 (▲) and 112 (+) kDa) was estimated from absorbance scans at 620 nm of the gels stained with Coomassie blue (left-hand panels). The P_{fr} λ_{max} determined from the difference spectrum (ΔA minimum) for each sample is also plotted as a function of time (right-hand panels). The first lane of each gel contains 1 μg of column-immonopurified phytochrome for comparison. (Data from Vierstra & Quail (1982 b).)

[44]

proteolytically cleaved fragment relative to the average composition of the remainder of the molecule. The consequences of these differences to the properties of the 6–10 kDa polypeptide fragment are unknown. In contrast to 118/114 kDa preparations, which have been reported to contain a mixture of two N-terminal amino acids, Lys and Ala (Hunt & Pratt 1980a), 124 kDa phytochrome has a blocked N-terminus (Vierstra & Quail 1983a).

The changes in properties that occur upon this adventitous proteolytic conversion of 124 kDa to 118/114 kDa phytochrome have provided valuable information on the molecular structure of the photoreceptor. It is clear that the 6–10 kDa polypeptide fragment(s) is critical to the structural and spectral integrity of the molecule and thus potentially important to its functional

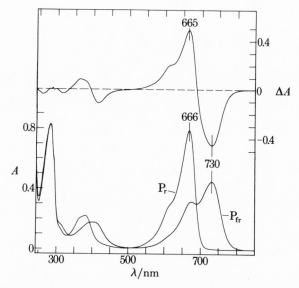

FIGURE 2. Absorbance and difference spectra of 124 kDa *Avena* phytochrome ($A_{666}/A_{280} = 0.97$). Absorbance spectra were measured after saturating red (P_{fr}) and far-red (P_r) irradiation, and the difference spectrum was obtained by subtracting the spectrum of P_{fr} from P_r. (Data from Vierstra & Quail (1983a).)

properties. The loss of the blocked N-terminus upon proteolysis indicates that at least part of this critical polypeptide domain is located at the N-terminus of the molecule. The differential susceptibility of the proteolytically sensitive site(s) in the P_r and P_{fr} forms indicates that phototransformation causes a change in molecular topology involving this domain. The report that the shift in P_{fr} λ_{max} with incubation as P_r in crude homogenates occurs in a wide range of species (Baron & Epel 1982) suggests a conservation of the site and its surrounding domain. The differential rates of bleaching of P_r and P_{fr} by tetranitromethane also reflect differences in chromophore chemistry and/or topology of the two forms.

AUTOREGULATION OF TRANSLATABLE mRNA LEVELS

A full understanding of the function of phytochrome in the living cell must include an understanding of the regulation of the levels of the photoreceptor. It has long been known that these levels are controlled both developmentally and by light. Phytochrome is synthesized *de novo* in the P_r form (Quail *et al.* 1973) accumulating in dark-grown tissue until a plateau is reached (Schäfer *et al.* 1972; Quail *et al.* 1973). This plateau represents a steady-state balance between the synthesis and degradation of P_r (Quail *et al.* 1973). Transfer of tissue to the light results in a rapid decline in phytochrome levels because P_{fr} has a much greater rate of degradation (100-fold in *Cucurbita*) than P_r (Quail *et al.* 1973; Schäfer *et al.* 1975; Hunt & Pratt 1979 *b*, 1980 *b*; Pratt 1979). In plants transferred to continuous white light a new constant level is reached at 1–3 % of the phytochrome initially present (Hunt & Pratt 1979 *b*; Jabben & Deitzer 1978), presumably representing a new steady-state balance between P_r synthesis and P_{fr} degradation. Return of light-treated tissue to the dark leads to a reaccumulation of phytochrome in the P_r form by continued synthesis *de novo* (Hunt & Pratt 1980 *b*; Quail *et al.* 1973; Schäfer *et al.* 1975).

These and other data have led to the generalized view of the phytochrome system *in vivo* shown in scheme 1, where ^0k_s is a zero-order rate constant of synthesis; $^1k_d^{P_r}$ and $^1k_d^{P_{fr}}$ are first-order rate constants of degradation of P_r and P_{fr} respectively, with $^1k_d^{P_r} \ll {}^1k_d^{P_{fr}}$; and P_r' and P_{fr}' are degradation products of P_r and P_{fr}. In this view phytochrome levels are considered to be modulated strictly at the protein level by the disparate rate constants of degradation for P_r and P_{fr} against a constant background rate of synthesis determined by an unchanging ^0k_s. Clearly implied in this scheme is a constant level of translatable phytochrome mRNA supporting a constant rate of *de novo* synthesis of the chromoprotein.

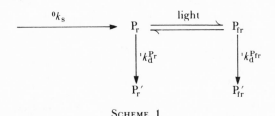

SCHEME 1

Recent studies on the cell-free synthesis of the phytochrome apoprotein have led us to consider a second level of control. Whereas translatable phytochrome mRNA is readily detectable in etiolated tissue, none is detectable in green tissue grown in continuous white light (Quail *et al.* 1983; Hershey *et al.* 1982; Gottmann & Schäfer 1982). When the levels of translatable phytochrome mRNA are monitored after irradiation of etiolated tissue with 5 s of red light, a very rapid decline in this translatable mRNA is observed (figure 3) (Colbert *et al.* 1983). The decline is detectable within 15–30 min, a 50 % reduction occurs within 50–60 min and more than 95 % reduction occurs within 2 h (figure 3 *b*). The effect of the red light pulse is reversed by an immediately subsequent far-red pulse to the level of the far-red light control, indicating that phytochrome exerts autoregulatory control over its translatable mRNA level (table 2). Red light dose–response curves show that the response is senstive to very low levels of P_{fr} (Colbert *et al.* 1983). Conversion of less than 1 % of the total cellular phytochrome to P_{fr} induces about 60 % of the maximum response, and 20 % P_{fr} saturates the response.

These data indicate that scheme 1 should be revised as shown in scheme 2 to reflect the dual control system involved. The photoconversion of phytochrome to P_{fr} not only enhances the degradation of the chromoprotein but also reduces its rate of synthesis by causing a decrease in the level of translatable phytochrome mRNA. This negative feedback modulation of translatable mRNA levels appears to be reversible upon depletion of P_{fr} in extended darkness. Gottmann & Schäfer (1982) have reported the reappearance of detectable levels of translatable phytochrome mRNA in green *Avena* returned to the dark for 10–24 h.

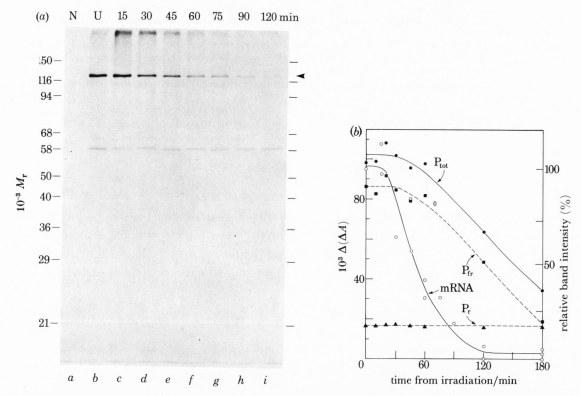

FIGURE 3. The effect of red irradiation on the levels of phytochrome and phytochrome mRNA in etiolated *Avena* shoots. (*a*) Time course of the decrease in translatable phytochrome mRNA in darkness after a 5 s saturating red irradiation. At the times indicated poly(A) RNA was isolated (Colbert *et al.* 1983) and translated (Pelham & Jackson 1976), and the phytochrome apoprotein (arrow) was immunoprecipitated (Ivarie & Jones 1979). The immunoprecipitates were subjected to sodium dodecyl sulphate polyacrylamide gel electrophoresis (Laemmli 1970) and fluorographed (Laskey 1980). Lane *A*, non-immune control (*N*), unirradiated tissue. Lanes *b–i*, antiphytochrome IgG immunoprecipitates. Lane *b*, unirradiated tissue (*U*). Lanes *c–i*, 5 s of red light followed by incubation in the dark for: lane *c*, 15 min; lane *d*, 30 min; lane *e*, 45 min; lane *f*, 60 min; lane *g*, 75 min; lane *h*, 90 min; lane *i*, 120 min. The minor band at *ca.* 58 kDa is thought to be a contaminant recognized by non-phytochrome antibodies in the antiphytochrome IgG preparations, because this band is not observed in immunoprecipitates from fractions containing phytochrome mRNA after size fractionation of poly(A) RNA on sucrose gradients (see figure 5). (*b*) Time-course of the change in cellular phytochrome and translatable phytochrome mRNA in etiolated shoots in darkness after a saturating red (5 s) irradiation. At the times indicated either poly(A) RNA was isolated and translated, or phytochrome was extracted and measured. The *in vitro* synthesized phytochrome apoprotein was immunoprecipitated, subjected to sodium dodecyl sulphate polyacrylamide gel electrophoresis, and quantitated by scanning fluorographs of these gels (○). Data points are from four different experiments including that in (*a*) and are expressed as a percentage of the initial levels in the unirradiated control in each case. Phytochrome ($10^3\Delta(\Delta A)$) was measured spectrophotometrically in crude extracts with $CaCO_3$ as a scattering agent. P_{tot}, total amount of spectrally detectable phytochrome (•); P_r (▲), P_{fr}(■), amounts of the two spectral forms that comprise P_{tot}. The levels of P_r and P_{fr} at time zero are those immediately after the 5 s red irradiation. Data points are the means of two different experiments in which each time point was duplicated. (Data from Colbert *et al.* (1983).)

There are several implications of these findings. First, whereas short-term fluctuations in phytochrome levels in response to light involve control predominantly at the protein level by the differential turnover rates of P_r and P_{fr}, in the long term the decreased rate of P_r synthesis needs to be accounted for. In particular, attempts to perform quantitative physiological studies under continuous irradiation or under diurnal light–dark cycles will need to take into consideration the expected fluctuating rates of synthesis as well. Second, the level of phytochrome measured in green tissue in continuous light (1–3 % of the etiolated tissue level (Hunt & Pratt

TABLE 2. AUTOREGULATORY CONTROL OF TRANSLATABLE PHYTOCHROME mRNA LEVELS

(Etiolated *Avena* seedlings were irradiated as indicated and returned to the dark for 3 h before poly(A) RNA was isolated and translated. Immunoprecipitates were prepared by using antiphytochrome IgG, subjected to sodium dodecyl sulphate polyacrylamide gel electrophoresis and fluorographed. The relative intensities of the phytochrome apoprotein band in each treatment were determined by scanning the fluorographs as in figure 3. Data from Colbert *et al.* (1983).)

treatment	translatable phytochrome mRNA level (relative units)
unirradiated control	100
5 s red + 3 h dark	5
5 s red + 6 s far red + 3 h dark	34
6 s far red + 3 h dark	37

1979 *b*; Jabben & Deitzer 1978)) (*a*) appears to be higher than expected of the combined effect of the more than 20-fold decrease in rate of synthesis (figure 3 *b*) and 100-fold increase in rate of degradation (Quail *et al.* 1973) that should apparently accompany P_{fr} formation, and (*b*) appears to be higher than expected given the fact that no phytochrome apoprotein is detectable

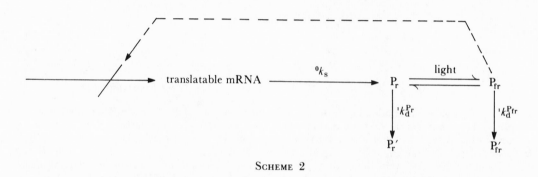

SCHEME 2

among the cell-free translation products of green-tissue poly(A) mRNA (Quail *et al.* 1983; Gottmann & Schäfer 1982). One potential explanaton of this apparent discrepancy that is being explored is that the phytochrome detected in green tissue is a second gene product, immunologically distinct from that predominating in etiolated tissue. Third, phytochrome mRNA (figure 3), together with protochlorophyllide reductase mRNA (Apel 1981), exhibits the most rapid change in phytochrome-regulated mRNA (15 min lag) so far reported. The rapidity of this autoregulation thus ironically makes phytochrome itself an attractive system with which to investigate the more general question of the mechanisms underlying phytochrome-regulated gene expression.

cDNA SYNTHESIS AND CLONING

We are in the process of cloning phytochrome cDNA for several purposes including the deduction of the primary sequence of the phytochrome polypeptide, the monitoring of potential developmental and light-induced changes in the physical abundance of phytochrome mRNA and the isolation and characterization of the phytochrome gene(s). This task has proved not to be simple because of the unexpectedly low abundance of the phytochrome mRNA. In contrast to the chromoprotein, which composes about 0.5 % of the soluble protein in etiolated *Avena* shoots (Bolton 1979; Vierstra & Quail 1983*a*), translatable phytochrome mRNA is less than 0.005 % of the poly(A) RNA in this tissue (Colbert *et al.* 1983; Hershey *et al.* 1982). By analogy with estimates for soybean tissue (Fischer & Goldberg 1982) this value corresponds to only about 25 phytochrome mRNA molecules per cell. Essentially all cDNA clones for identified plant nuclear gene products thus far isolated correspond to mRNAs that are in the medium to high abundance category (2–50 % of the cellular mRNA mass) (see, for example, Bedbrook *et al.* 1980; Bernal-Lugo *et al.* 1981; Broglie *et al.* 1981; Chandler *et al.* 1983; Geraghty *et al.* 1981; Meinke *et al.* 1981). Moreover, although there is an increasing number of examples from mammalian systems where cDNAs for low abundance mRNAs have been isolated (Edge & Markum 1982; Korman *et al.* 1982; Kraus & Rosenberg 1982; Parnes *et al.* 1981; Sood *et al.* 1981), these are still relatively few and represent mRNAs that are 0.01–0.2 % of the total cellular mRNA, above that of phytochrome mRNA.

Our initial attempts to overcome this handicap have involved exploiting the relatively large size of the phytochrome mRNA to provide enrichment by size fractionation and using a synthetic oligodeoxynucleotide complementary to part of the mRNA sequence as a hybridization probe for screening a cDNA library for phytochrome clones. Poly(A) RNA from etiolated *Avena* tissue has been fractionated by rate-zonal sucrose gradient centrifugation, and the fractions containing phytochrome mRNA have been located by *in vitro* translation and immunoprecipitation (figure 4). This procedure provides an enrichment of 25–40-fold in a single pass and a further 4–5-fold by a second gradient. Although phytochrome mRNA still composes only 1.0–1.5 % of the mRNAs in the most enriched fraction, it has been used to synthesize cDNA (figure 5), the double-stranded cDNA has been inserted into the *Hin*d III site of pBR322 by using synthetic linkers, and the recombinant plasmids have been used to transform *Escherichia coli* RR1.

The successful use of synthetic oligonucleotide probes to isolate cDNA clones representing the low-abundance interferon mRNA (Edge & Markum 1982) encouraged us to adopt this strategy as the most direct and efficient means of identifying phytochrome cDNA clones. The sequence of the probe used was determined from the only phytochrome protein sequence available, a segment of 11 amino acids associated with the chromophore (figure 6). Because of the limited extent of the known polypeptide sequence and the generally high degree of degeneracy in the corresponding nucleotide sequence, it was necessary to synthesize a wobble mix of 24 different 14-mers to accommodate all possible permutations.

Initial attempts to use the ^{32}P-labelled 14-mers directly as colony hybridization probes failed to identify colonies hybridizing strongly above background, possibly because the average G–C content of the wobble mixture is only 33 %. The differential between the stringency conditions permitting specific and non-specific duplex formation is expected to be very narrow for this oligonucleotide. It was thus decided to generate longer probes of much higher specific activity by using the 14-mers as selective primers for cDNA synthesis directed by reverse transcriptase with poly(A) RNA enriched for phytochrome mRNA as a template. Some 24 positively hybridizing colonies were isolated from the cDNA library in this fashion and subjected to a Northern blot secondary screening procedure. Adjacent tracks of poly(A) RNA – one from unirradiated tissue and one from tissue exposed to light to deplete the phytochrome mRNA – were probed with nick-translated plasmids from the various positive clones. Plasmids carrying phytochrome inserts would be expected to hybridize to an mRNA that is depleted in the poly(A) fraction from light-exposed tissue (assuming transcriptional control) and that migrates as a 3.5–4 kilobase species. No clone with both the required characteristics was located.

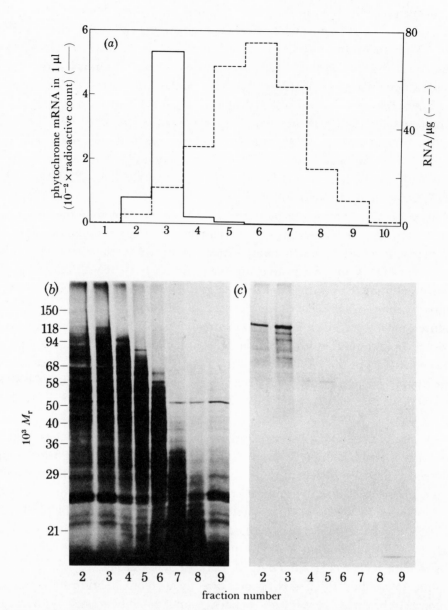

FIGURE 4. Fractionation of poly(A) RNA from etiolated *Avena* tissue. Poly(A) RNA was enriched for phytochrome mRNA by centrifugation through a linear sucrose gradient (50–300 g l⁻¹). After fractionation of the gradient, mRNA from each fraction was recovered and translated in an mRNA-dependent reticulocyte lysate. (*a*) –––, Distribution of RNA in the gradient as determined by A_{260}; —, distribution of phytochrome mRNA in the gradient as determined by translation and immunoprecipitation with antiphytochrome IgG. (*b*) Sodium dodecyl sulphate polyacrylamide gel electrophoresis and fluorography of translation products synthesized in response to gradient fractionated mRNA by a nuclease-inactivated reticulocyte lysate. Equal amounts of radioactivity were loaded on each lane. (*c*) Sodium dodecyl sulphate polyacrylamide gel electrophoresis of products immunoprecipitated by antiphytochrome IgG from the total translation products of the gradient fractions shown in (*b*). Fractions enriched for translatable phytochrome mRNA have been recentrifuged on a linear sucrose gradient (150–300 g l⁻¹) to further enrich the phytochrome mRNA. The phytochrome mRNA in the peak fraction of this second gradient composed about 1 % of the poly(A) in this fraction, thus representing about a 200-fold enrichment relative to the initial poly(A) fraction. (Data from Quail *et al.* (1983).)

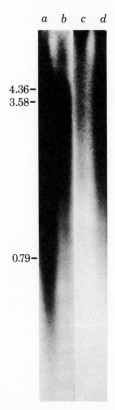

FIGURE 5. cDNA synthesized from the fraction most enriched in phytochrome mRNA from the second pass sucrose gradient referred to in figure 4. High molecular mass double-stranded (ds) cDNA was synthesized either by a modification of the method of Wickens *et al.* (1978) or by a modification of the method of Land *et al.* (1981). The size of cDNA made by each procedure was measured on 1.5% (by mass) agarose gels. In both cases the cDNA averaged over 2.8 kilobase pairs (lane *a*, Land *et al.*; lane *b*, Wickens *et al.*), with both containing cDNA corresponding to the upper molecular mass limit of the RNA used to direct first strand synthesis. ds cDNA was more than 95% S_1-resistant (lanes *c* and *d*). ^{32}P-labelled *Hind* III linkers were blunt-end ligated to the cDNA shown in lane *b*. The cDNA was digested with *Hind* III, separated from the linkers, and ligated into phosphatase-treated *Hind* III digested pBR 322. The numbers at the left are kilobase pairs.

A possible reason for these results is that the temperature and salt conditions required for the reverse transcriptase reaction cause non-specific priming of cDNA synthesis. Oligonucleotides specific for only one mRNA have been shown in other systems to prime cDNA synthesis from a number of different mRNA species (Edge & Markum 1982). Our positive clones were found to be complementary to mRNAs of various sizes from *ca.* 1 to 7.5 kilobases, indicating a high level of this non-specific priming, probably exacerbated by the A–T-rich nature of the oligonucleotide.

The difficulties experienced with the synthetic probe and the observation made during this work that translatable phytochrome mRNA rapidly declines to less than one-twentieth of its original level upon P_{fr} formation has led us recently to explore a more conventional 'dark versus light' difference screening procedure. This approach has been used successfully to isolate cDNA clones for ribulose bisphosphate carboxylase small subunit (Bedbrook *et al.* 1980), chlorophyll *a/b* protein (Apel 1982) and protochlorophyllide reductase (Apel 1982). The strategy is to screen replica arrays of colonies with probes constructed separately from mRNA isolated from dark- and from light-treated tissue. Colonies displaying different degrees of hybridization with

(a)

chromophore
|

protein N –	Leu	Arg	Ala	Pro	His	Ser	Cys	His	Leu	Gln	Tyr	– C
mRNA 5′ –	CUU	CGU	GCU	CCU	CAU	UCU	UGU	CAU	CUU	CAA	UAU – 3′	
	CUC	CGC	GCC	CCC	CAC	UCC	UGC	CAC	CUC	CAG	UAC	
	CUA	CGA	GCA	CCA		UCA			CUA			
	CUG	CGG	GCG	CCG		UCG			CUG			
		AGA				AGU			UUA			
		AGG				AGC			UUG			

(b)

protein N –	Cys	His	Leu	Gln	Tyr	– C
mRNA 5′ –	UGU	CAU	CUU	CAA	UAU – 3′	
	UGC	CAC	CUC	CAG	UAG	
			CUA			
			CUG			
			UUA			
			UUG			

$$\text{oligonucleotide 14-mer (+) strand} \qquad 3' - \ AC{}^{A}_{G} \ \ GT{}^{A}_{G} \ \left[\begin{array}{c} GA{}^{T}_{C} \\ G \\ AA{}^{T}_{C} \end{array}\right] \ GTT \ \ AT \qquad - 5'$$

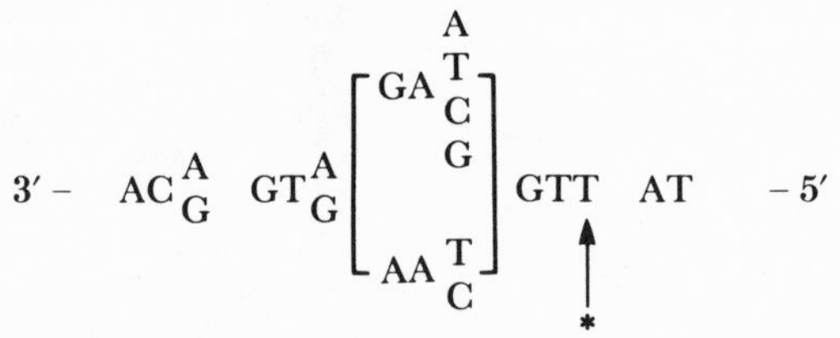

FIGURE 6. Known amino acid sequence adjacent to the phytochrome chromophore (Fry & Mumford 1971; Lagarias & Rapoport 1980) and the corresponding mRNA sequence used to design a synthetic oligonucleotide probe. (a) Entire known sequence. (b) Carboxy-terminus sequence used to specify synthetic oligonucleotide sequence. This region exhibits the least degeneracy available in the sequence. The 24 different 14-mer sequences synthesized in the wobble mix are indicated at the bottom. The arrow indicates the single potential mismatch in the sequence where only T was used instead of the prescribed T/C ambiguity.

the two probes are tentatively identified as light-regulated. Although this approach relies on the as-yet untested assumption that the mass rather than the translatability of phytochrome mRNA is controlled, precedent with other light-regulated gene products indicates that this is a reasonable first assumption (Apel 1982; Bedbrook et al. 1980; Everett et al. 1982). Figures 7 and 8 show the relative levels of translatable phytochrome mRNA in gradient fractionated poly(A) RNA from unirradiated and from irradiated tissues. We have constructed twin cDNA probes from the gradient fractions enriched (figure 7) and depleted (figure 8) in phytochrome mRNA and are currently screening the cDNA library.

CONCLUSIONS

Our concepts of the phytochrome molecule and the regulation of its levels *in vivo* are currently undergoing rapid revision. Investigation of the properties of the newly purified 124 kDa phytochrome will, it is hoped, yield an accurate picture of the functional properties of the native photoreceptor. Moreover, whereas at first it may be tempting to dismiss data obtained with the 118/114 kDa phytochrome as obsolete, experience so far indicates that judicious comparisons with the 124 kDa molecule can be expected to provide further valuable insights into the molecular structure of the chromoprotein. The discovery that phytochrome exerts negative feedback control over its own mRNA levels is interesting teleologically and has unexpectedly

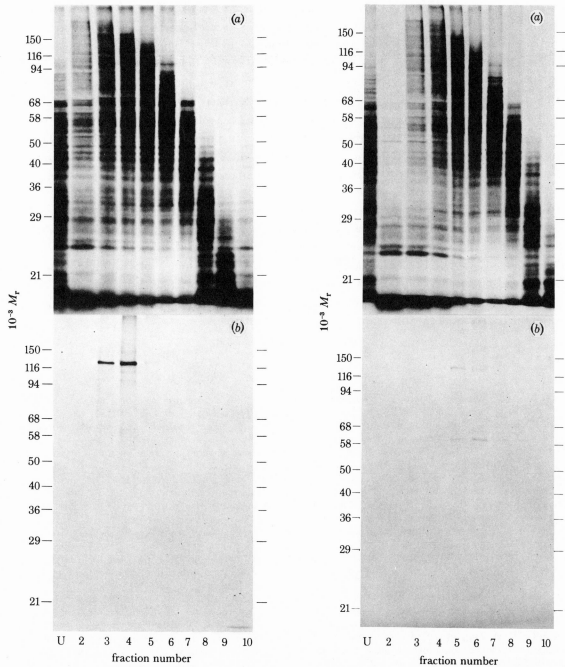

FIGURE 7. Sucrose gradient (50–300 g l⁻¹) fractionation of polysomal poly(A) RNA from etiolated, unirradiated *Avena* shoots. (*a*) Total translation products from each fraction. (*b*) Immunoprecipitates prepared from each fraction with antiphytochrome IgG. Phytochrome mRNA, concentrated in fractions 3 and 4, is enriched 10–20-fold relative to initial unfractionated poly(A) RNA.

FIGURE 8. Sucrose gradient (50–300 g l⁻¹) fractionation of polysomal poly(A) RNA from etiolated oats irradiated with 5 min of red light and returned to the dark for a further 3 h before harvest. (*a*) Total translation products from each fraction. (*b*) Immunoprecipitates prepared from each fraction with antiphytochrome IgG. Phytochrome mRNA is in fractions 5 and 6 at a considerably lower level than in comparable fractions from unirradiated tissue (figure 7).

[53]

provided an attractive system for investigating phytochrome-regulated gene expression. It seems reasonable that plants in the light may no longer need to synthesize the photoreceptor at the rates prevailing in seedlings before their first exposure to light. The relatively high levels of the photoreceptor observed in dark-grown plants may serve a light-scavenging function in the pre-emergent seedling, a function no longer needed after de-etiolation. The apparent retention in light-grown plants of the capacity to reinstate higher rates of synthesis upon P_{fr} depletion in prolonged darkness or under shaded conditions may serve a role in the photoperiod-monitoring and shade-detecting systems of plants. Experimentally, the rapid autoregulation of translatable phytochrome mRNA permits attention to be focused on a relatively short period in attempts to define the sequence of events between signal perception and altered gene expression.

We thank L. H. Pratt for initial assistance in producing antiphytochrome IgG, W. R. Briggs in whose laboratory the antiphytochrome IgG was produced and M. G. Murray for valuable discussions. The work was supported in part by National Science Foundation grants PCM 8003921, PCM 7723584, PCM 7514161, Department of Energy grant no. DE-AC02-81ER10903 and U.S. Department of Agriculture, Science and Education Administration grant no. 59-2551-1-1-744-0. J.T.C. was supported as a predoctoral trainee by National Institutes of Health grant no. 5T32GM07215.

REFERENCES

Apel, K. 1981 The protochlorophyllide holochrome of barley. Phytochrome-induced decrease of translatable mRNA coding for the NADPH-protochlorophyllide oxidoreductase. *Eur. J. Biochem.* **120**, 89–93.

Apel, K. 1982 The light-dependent control of plastid development in barley. In *Structure and function of plant genomes* (Abstracts of NATO Conference, Porto Potesa, Italy, 23 August–2 September), p. 23.

Baron, O. & Epel, B. L. 1982 Studies on the capacity of P_r *in vitro* to photoconvert to the long-wavelength Pfr form. A survey of ten plant species. *Photochem. Photobiol.* **36**, 79–82.

Bedbrook, J. R., Smith, S. M. & Ellis, R. J. 1980 Molecular cloning and sequencing of cDNA encoding the precursor to the small subunit of chloroplast ribulose-1,5-bisphospate carboxylase. *Nature, Lond.* **287**, 692–697.

Bernal-Lugo, I., Beachy, R. N. & Varner, J. E. 1981 The response of barley aleurone layers to gibberellic acid includes the transcription of new sequences. *Biochem. biophys. Res. Commun.* **102**, 617–623.

Bolton, G. W. 1979 Phytochrome: aspects of its protein and photochemical properties. Ph.D. thesis, University of Minnesota.

Bolton, G. W. & Quail, P. H. 1982 Cell-free synthesis of phytochrome apoprotein. *Planta* **155**, 212–217.

Briggs, W. R. & Rice, H. V. 1972 Phytochrome: Chemical and physical properties and mechanism of action. *A. Rev. Pl. Physiol.* **23**, 293–334.

Broglie, R., Bellemare, G., Bartlett, S., Chua, N. H. & Cashmore, A. R. 1981 Cloned DNA sequences complementary to mRNAs encoding precursors to the small subunit of ribulose-1,5-bisphosphate carboxylase and a chlorophyll *a/b* binding polypeptide. *Proc. natn. Acad. Sci. U.S.A.* **78**, 7304–7308.

Chandler, P. M., Higgins, T. J. V., Randall, P. J. & Spencer, D. 1983 Regulation of legumin levels in developing pea seeds under conditions of sulfur deficiency. Rates of legumin synthesis and levels of legumin mRNA. *Pl. Physiol.* **71**, 47–54.

Colbert, J. T., Hershey, H. P. & Quail, P. H. 1983 Autoregulatory control of translatable phytochrome mRNA levels. *Proc. natn. Acad. Sci. U.S.A.* **80**, 2248–2252.

Edge, M. D. & Markham, A. F. 1982 Applications of oligonucleotide synthesis to interferon research. *Biochim. biophys. Acta* **695**, 35–48.

Epel, B. L. 1981 A partial characterization of the long-wavelength 'activated' far-red absorbing form of phytochrome. *Planta* **151**, 1–5.

Everett, M. S. & Briggs, W. R. 1970 Some spectral properties of pea phytochrome *in vivo* and *in vitro*. *Pl. Physiol.* **45**, 679–683.

Everett, M., Polans, N., Jorgensen, R. A., Palmer, J. D. & Thompson, W. F. 1982 Phytochrome regulation of cytoplasmic and chloroplast RNA levels. In *Structure and function of plant genomes* (abstracts of NATO Conference, Porto Portesa, Italy, 23 August–2 September), p. 64.

Fischer, R. L. & Goldberg, R. B. 1982 Structure and flanking regions of soybean seed protein genes. *Cell* **29**, 651–660.

Fry, K. T. & Mumford, F. E. 1971 Isolation and partial characterization of a chromophore-peptide fragment from pepsin digests of phytochrome. *Biochem. biophys. Res. Commun.* **45**, 1466–1473.

Gardner, G., Pike, C. S., Rice, H. V. & Briggs, W. R. 1971 'Disaggregation' of phytochrome *in vitro* – a consequence of proteolysis. *Pl. Physiol.* **48**, 686–693.

Geraghty, D., Peifer, M. A., Rubenstein, I. & Messing, J. 1981 The primary structure of a plant storage protein: zein. *Nucl. Acids Res.* **9**, 5163–5174.

Gottman, K. & Schäfer, E. 1982 *In vitro* synthesis of phytochrome apoprotein directed by mRNA from light and dark grown *Avena* seedlings. *Photochem. Photobiol.* **35**, 521–525.

Hahn, T. R., Kang, S. S. & Song, P. S. 1980 Difference in the degree of exposure of chromophores in the Pr and Pfr forms of phytochrome. *Biochem. biophys. Res. Commun.* **97**, 1317–1323.

Hahn, T. R. & Song, P. S. 1981 Hydrophobic properties of phytochrome as probed by 8-anilinonaphthalene-1-sulfonate fluorescence. *Biochemistry, Wash.* **20**, 2602–2609.

Hahn, T. R., Song, P. S., Quail, P. H. & Vierstra, R. D. 1983 Tetranitromethane oxidation of phytochrome chromophore as a function of spectral forum and molecular weight. *Pl. Physiol.*, submitted.

Hershey, H. P., Colbert, J. T., Murray, M. G. & Quail, P. H. 1982 Phytochrome mRNA: autoregulation of levels *in vivo*, partial purification and cDNA synthesis. In *Structure and function of plant genomes* (Abstracts of NATO Conference, Port Portese, Italy, 23 August–2 September), p. 65.

Hunt, R. E. & Pratt, L. H. 1979a Phytochrome immunoaffinity purification. *Pl. Physiol.* **64**, 332–336.

Hunt, R. E. & Pratt, L. H. 1979b Phytochrome radioimmunoassay. *Pl. Physiol.* **64**, 327–331.

Hunt, R. E. & Pratt, L. H. 1980a Partial characterization of undegraded oat phytochrome. *Biochemistry, Wash.* **19**, 390–394.

Hunt, R. E. & Pratt, L. H. 1980b Radioimmunoassay of phytochrome content in green, light-grown oats. *Pl. Cell Envir.* **3**, 91–95.

Ivarie, R. D. & Jones, P. P. 1979 A rapid sensitive assay for specific protein translations: use of *Staphylococcus aureus* as an adsorbent for immune complexes. *Analyt. Biochem.* **97**, 24–35.

Jabben, M. & Deitzer, G. F. 1978 Spectrophotometric phytochrome measurements in light-grown *Avena sativa* L. *Planta* **143**, 309–313.

Kerscher, L. & Nowitzki, S. 1982 Western blot analysis of a lytic process *in vitro* specific for the red light absorbing form of phytochrome. *FEBS Lett.* **146**, 173–176.

Korman, A. J., Knudsen, P. J., Kaufman, J. F. & Strominger, J. L. 1982 cDNA clones for the heavy chain of HLA-DR antigens obtained after immunopurification of polysomes by monoclonal antibody. *Proc. natn. Acad. Sci. U.S.A.* **79**, 1844–1848.

Kraus, J. P. & Rosenberg, L. E. 1982 Purification of low abundance messenger RNAs from rat liver by polysome immunoadsorption. *Proc. natn. Acad. Sci. U.S.A.* **79**, 4015–4019.

Laemmli, U. K. 1970 Cleavage of structural proteins during the assembly of the head of bacteriophage T_4. *Nature, Lond.* **227**, 680–685.

Lagarias, J. C. & Rapoport, H. 1980 Chromopeptides from phytochrome. The structure and linkage of the Pr form of the phytochrome chromophore. *J. Am. chem. Soc.* **102**, 4821–4828.

Land, H., Grez, M., Hanser, H., Lindenmaier, W. & Schütz, G. 1981 5'-terminal sequences of eucaryotic mRNA can be cloned with high efficiency. *Nucl. Acids Res.* **9**, 2251–2266.

Laskey, R. A. 1980 The use of intensifying screens or organic scintillators for visualizing radioactive molecules resolved by gel electrophoresis. *Methods Enzymol.* **65**, 363–371.

Meinke, D. W., Chen, J. & Beachy, R. N. 1981 Expression of storage protein genes during soybean seed development. *Planta* **153**, 130–139.

Mumford, F. E. & Jenner, E. L. 1966 Purification and characterization of phytochrome from oat seedlings. *Biochemistry, Wash.* **5**, 3657–3662.

Parnes, J. R., Baruch, V., Felsenfeld, A., Ramanathan, L., Ferini, V., Appella, E. & Seidman, J. G. 1981 Mouse β_2-microglobulin cDNA clones: a screening procedure for cDNA clones corresponding to rare mRNAs. *Proc. natn. Acad. Sci. U.S.A.* **78**, 2253–2257.

Pelham, H. R. B. & Jackson, R. J. 1976 An efficient mRNA-dependent translation system from reticulocyte lysates. *Eur. J. Biochem.* **67**, 247–256.

Pike, C. S. & Briggs, W. R. 1972 The dark reactions of rye phytochrome *in vivo* and *in vitro*. *Pl. Physiol.* **49**, 514–520.

Pratt, L. H. 1975 Photochemistry of high molecular weight phytochrome *in vitro*. *Photochem. Photobiol.* **22**, 33–36.

Pratt, L. H. 1979 Phytochrome: function and properties. *Photochem. Photobiol. Rev.* **4**, 59–124.

Pratt, L. H. 1982 Phytochrome: the protein moiety. *A. Rev. Pl. Physiol.* **33**, 557–582.

Pratt, L. H. & Cundiff, S. C. 1975 Spectral characterization of high-molecular weight phytochrome. *Photochem. Photobiol.* **21**, 91–97.

Quail, P. H., Bolton, G. W., Hershey, H. P. & Vierstra, R. D. 1983 Phytochrome: molecular weight, *in vitro* translation and cDNA cloning. In *Current topics in plant biochemistry–physiology* (ed. D. Randall, D. G. Blevins & R. Larson), Columbia, Missouri: University of Missouri, pp. 25–36.

Quail, P. H., Bolton, G. W. & Vierstra, R. D. 1981 Molecular properties of phytochrome synthesized *in vitro* and extracted from etiolated tissue. In *Light and the expression of genes in plants and microorganisms* (Abstracts, University of Hanover Symposium, 1–3 July), p. 2.

[55]

Quail, P. H., Schäfer, E. & Marmé, D. 1973 Turnover of phytochrome in pumpkin cotyledons. *Pl. Physiol.* **52**, 128–131.

Rice, H. V. & Briggs, W. R. 1973 Partial characterization of oat and rye phytochrome. *Pl. Physiol.* **51**, 927–938.

Rice, H. V., Briggs, W. R. & Jackson-White, C. J. 1973 Purification of oat and rye phytochrome. *Pl. Physiol.* **51**, 917–926.

Roux, S. J., McEntire, K. & Brown, W. E. 1982 Determination of extinction coefficients of oat phytochrome by quntitative amino acid analysis. *Photochem. Photobiol.* **35**, 537–543.

Sarkar, H. K. & Song, P. S. 1982 Nature of phototransformation of phytochrome as probed by intrinsic tryptophan residues. *Biochemistry, Wash.* **21**, 1967–1972.

Schäfer, E., Lassig, T. U. & Schopfer, P. 1975 Photocontrol of phytochrome destruction in grass seedlings. The influence of wavelength and irradiance. *Photochem. Photobiol.* **22**, 193–202.

Schäfer, E., Marchal, B. & Marmé, D. 1972 *In vivo* measurements of the phytochrome photostationary state in far red light. *Photochem. Photobiol.* **15**, 457–464.

Song, P. S., Chae, Q. & Gardner, J. D. 1979 Spectroscopic properties and chromophore conformations of the photomorphogenic receptor phytochrome. *Biochim. biophys. Acta* **576**, 479–495.

Sood, A. K., Pereira, D. & Weissman, S. M. 1981 Isolation and partial nucleotide sequence of a cDNA clone for human histocompatibility antigen HLB-A by use of an oligodeoxynucleotide primer. *Proc. natn. Acad. Sci. U.S.A.* **78**, 616–620.

Tobin, E. M. & Briggs, W. R. 1973 Studies on the protein conformation of phytochrome. *Photochem. Photobiol.* **18**, 487–495.

Vierstra, R. D. & Quail, P. H. 1982a Native phytochrome: inhibition of proteolysis yields a homogeneous monomer of 124 kilodaltons from *Avena*. *Proc. natn. Acad. Sci. U.S.A.* **79**, 5272–5276.

Vierstra, R. D. & Quail, P. H. 1982b Proteolysis alters the spectral properties of 124 kdalton phytochrome from *Avena*. *Planta* **156**, 158–165.

Vierstra, R. D. & Quail, P. H. 1983a Purification and initial characterization of 124 kdalton phytochrome from *Avena*. *Biochemistry, Wash.* **22**, 2498–2505.

Vierstra, R. D. & Quail, P. H. 1983b Photochemistry of 124 kdalton *Avena* phytochrome *in vitro*. *Pl. Physiol.* **72**, 264–267.

Wickens, M. P., Buell, G. N. & Schimke, R. T. 1978 Synthesis of double-stranded DNA complementary to lysozyme, ovomucoid and ovalbumin mRNAs. Optimization for full length second strand synthesis by *Escherichia coli* DNA polymerase I. *J. biol. Chem.* **253**, 2483–2495.

Discussion

W. Rüdiger (*Botanisches Institut der Universität München, F.R.G.*). The absorption maximum of P_{fr} has been reported to be at about 730 nm for native phytochrome but at about 720 nm for all forms of proteolytically degraded phytochrome. We found, however, that the absorption maximum of degraded phytochrome shifts from 720 nm (room temperature) to 730 nm at low temperature, e.g. at $-40\ ^\circ$C. The influence of the small part of the peptide chain that is split off by the first proteolysis can therefore be described in the first approximation as 'freezing' of the chromophore.

Phil. Trans. R. Soc. Lond. B **303**, 403–417 (1983)

Printed in Great Britain

Do plant photoreceptors act at the membrane level?

By J. A. Raven

Department of Biological Sciences, University of Dundee, Dundee DD1 4HN, U.K.

All of the photoreceptors involved in the absorption and transduction of light energy in photosynthesis are integral (carotenoid, chlorophyll) or peripheral (phycobilin) membrane proteins. The informational photoreceptors (phytochrome) and the flavoprotein (carotenoprotein?) cryptochrome, could be integral (carotenoprotein, flavoprotein) or peripheral or soluble (phytochrome, flavoprotein) pigment–protein complexes. The primary activity of the informational photoreceptors is unlikely to involve energization of primary active transport: the solute fluxes produced in this way would not form a quantitatively significant link in the perception–transduction–response sequence. By contrast, regulation of mediated solute fluxes at the plasmalemma could effect a substantial amplification of the absorbed photon signal, i.e. a large change in moles of solute transported could result from the absorption of 1 mol of photons. Modulation of the passive influx (or active efflux) of protons or calcium ions at the plasmalemma are likely targets for regulation by photoreceptors. Calcium flux regulation is particularly attractive in view of the ubiquity of calmodulin activity in eukaryotes, although problems could arise in maintaining the uniqueness of phytochrome messages *vis-à-vis* cryptochrome messages. Temporal analysis of the relation between photoreceptor changes and electrical effects resulting from changes in ion fluxes cannot, in general, rule out the involvement of intermediates between the redox or conformational change in the photoreceptor and the observed change in ion flux. Although slow in terms of the potential rate of change on solute fluxes resulting from direct interaction of a photoreceptor and a solute porter, the observed rates of signal transduction are well in excess of any obvious 'need' on the part of the plant in terms of rates of response to environmental changes.

1. Introduction

The explicit or implicit hypothesis underlying much recent work on photoperception by plants requires that membranes are involved at an early stage in the perception–transduction–response sequence (Marmé 1977; Raven 1981; Senger 1980). It is to these stages in photoperception that this paper is addressed, with particular emphasis on the possibility that transmembrane fluxes of solutes are early and essential events in the photoperception process. The analysis of the close temporal (and spatial?) coupling of light absorption and solute transport requires some discussion of both the nature of the photoreceptors and the sorts of transport systems with which they could interact. Particular emphasis will be placed on the distinction between direct (within a protein, or protein–protein) and indirect interaction in coupling photon absorption to changes in solute fluxes, and to the distinction between energetic and informational coupling of light absorption to solute fluxes. In addition to the permissible 'if' and 'how' of membrane involvement, the paper finally addresses the less widely acceptable question of *why* membranes are involved in photoperception in plants.

The discussion will centre mainly on oxygen-evolving photolithotrophs, and on the role of pigments other than the main photosynthetic pigments; however, these restrictions will not be rigidly adhered to when important results or hypotheses have originated from work on other organisms.

TABLE 1. SOME PROPERTIES OF PIGMENT–PROTEIN COMPLEXES THAT MAY BE INVOLVED IN PHOTOPERCEPTION IN PLANTS

chromophore	absorption maxima†/nm	specific absorption coefficient $m^3\,mol^{-1}\,m^{-1}$	lifetime of excited singlet state/s	photoredox reactions in vivo	light-powered ion pump in vivo	antenna role in vivo	light-induced conformational change in vivo	references
carotenoids	445–515	1.3×10^4 (β-carotene, 451 nm)	10^{-15}	no		yes	(yes)	Shropshire (1980), Song (1980)
retinol[1]	500–550	4×10^3 (500 nm)		no	yes: H^+/Cl^-	?	yes	Birge (1981), Shropshire (1980), Schobert & Lanyi (1982)
flavins	450	1.5×10^3 (450 nm)	$0.6\text{–}6.0 \times 10^{-9}$	yes	(no?)	no	yes	Raven (1981), Schmid (1980), Song (1980)
chlorophylls[2] (and other Mg-porphyrins)	430–450; 630–700 (chlorophylls a, b, c_1, c_2)	$1.0\text{–}1.3 \times 10^4$ (430–450 nm); $0.12\text{–}0.80 \times 10^4$ (620–700 nm)	$0.5\text{–}5.0 \times 10^{-9}$	yes	no	yes	yes	Meeks (1974), Nobel (1974), Giaquinta et al. (1975)
Fe porphyrins	415, 553, 522 (reduced Cyt c_{553} of Petalonia fascia)	2×10^4 (reduced Cyt c_{553} of Petalonia fascia at 415 nm)		yes[3] (phototaxin)	(no)	no	(yes)	Lemberg (1975), Yakushiji (1971), Poff et al. (1974)
phycobilin chromophore	540 (phycoerythrin); 625 (phycocyanin)	1.5×10^4 (phycoerythrin, 540 nm); 1.25×10^4 (phycocyanin, 625 nm)	$0.1\text{–}1.0 \times 10^{-9}$	no	no	yes	(yes)	Gantt (1981), Nobel (1974)
phytochrome chromophore	660 (P_r); 730 (P_{fr})	10^4 (P_r at 660 nm; P_{fr} at 730 nm)		no	no	?[4]	yes, in $P_r \to P_{fr}$, and $P_{fr} \to P_r$	Smith (1975), Raven (1981)

The data are mainly taken from pigments found in oxygen-evolving photolithotrophs; exceptions are denoted by superscript numbers: [1] rhodopsin found in metazoan photoreceptors; bacteriorhodopsin, halorhodopsin in the bacterial genus *Halobacterium*; [2] phototaxin occurs in the slime-mould *Dictyostelium*; [3] bacteriochlorophylls in photosynthetic bacteria; [4] the likelihood of such an antenna role for phytochrome *in vivo* is greatly decreased by the absence of a spectroscopically detectable reaction partner to which excitation energy could be transferred (W. Rüdiger, personal communication).
† In decreasing order of absorbance.

TABLE 2. INTRACELLULAR LOCATION OF PIGMENT–PROTEIN COMPLEXES IN PLANTS

pigment–protein complex	soluble in N phase	soluble in P phase	peripheral on N side of membrane	peripheral on P side of membrane	integral in membrane	references
carotenoproteins	no	no	(no) (light-harvesting peridinin-chlorophyll a complex of thylakoid membranes of Dinophyceae?)	no	light-harvesting and reaction centre chlorophyll–carotenoid complexes of thylakoid membranes	Song (1980), Prezelin (1981), Anderson et al. (1982)
retinol–opsins	no	no	no	no	rhodopsin in metazoan photoreceptor membrane; bacteriorhodopsin, halorhodopsin in Halobacterium plasmalemma	Birge (1981)
flavoproteins	nitrate reductase (cytosol)	glycollate oxidase (microbodies)	flavodoxin (ferredoxin replacement in some algae)	no	NADH–UQ, succinate–UQ oxidoreductase of inner mitochondrial membrane; ferredoxin–NADP$^+$ oxidoreductase in thylakoids (N side); flavoprotein–cytochrome b complex of Zea, Neurospora plasmalemma	Ragan (1976), Trebst (1974), Caubergs et al. (1983), Tolbert (1981), Husain et al. (1976), Ninnemann & Klemme-Wolframm (1976)
chlorophylls	no	no	(no) (see entry for carotenoproteins)	no	yes: see entry for carotenoproteins	see entry for carotenoproteins
Fe-porphyrins	phototaxin? (cytosol) catalase (plastid stroma)	catalase (microbodies)	?	c-type cytochromes in algal thylakoids, between mitochondrial membranes	cytochrome b–c_1 and a–a_3 complexes in inner mitochondrial membrane; cytochrome b–f complex in thylakoid membrane, Cyt b in plasmalemma	Hauska et al. (1982), Wikstrom et al. (1981), Caubergs et al. (1983), Tolbert (1981), Poff et al. (1974), Halliwell (1981)
photosynthetic phycobilins	no	inside thylakoid of Cryptophyceae	outside thylakoid of Cyanobacteria, Rhodophyceae	no	no	Gantt (1981), MacColl & Burns (1978)
phytochrome	yes (cytosol)	no	(yes)	no	no	Smith (1975), Haupt (1982), Marmé (1977)

2. Properties and location of photoreceptors

Almost all of the protein-associated chromophores that have visible absorption bands have been implicated at some time or other in the perception of light by plants. Table 1 shows some of the characteristics of pigment–protein complexes that are known to be, or may be, involved in photoperception. All of the chromophores have high specific absorption coefficients, thus making them effective photon absorbers at their respective absorption maxima. The lifetime of the excited singlet state varies with the environment of the chromophore, but is uniformly short for carotenoid and retinol pigments; despite this, the carotenoids can act as photo-sensitizers (Song 1980). The carotenoids and phycobilins seem to have no usable photoredox activities, unlike the rest of the pigments listed in table 1. The retinol–protein complexes are distinguished by their capacity to carry out (when membrane-associated) active proton trans-port apparently unrelated to internal redox reactions. Conformational changes related to photon absorption are widespread among the pigment–protein complexes.

Table 2 gives some information on the location of the major classes of pigment–protein complexes in cells, distinguishing between 'soluble', 'peripheral to membrane' and 'integral in membrane', as well as differentiating between the N and the P sides of the membranes (Singer 1974; Mitchell 1979). The carotenoproteins and retinoproteins are all integral mem-brane proteins, as are the Mg porphyrin-proteins; the flavoproteins and Fe porphyrin-proteins have a more catholic distribution, with different representatives of the two classes occurring as integral, peripheral and soluble pigment–protein complexes. The phycobilins, and phyto-chrome, are peripheral or soluble; the categories are not rigid, an example being the occurrence of phytochrome as both a soluble and a membrane-peripheral entity.

We shall see in § 3 that some integral membrane proteins can function as catalysts of solute transport across the membrane in which they occur: if these proteins are associated with a chromophore, then photon absorption by the chromophore *could* give a direct informational or energetic coupling to transport within a single polypeptide. In all other cases, a less direct coupling between photon absorption and solute transport must be envisaged, involving the direct transfer of excitation energy or conformational energy from the photoreceptor protein to the porter (by protein–protein interaction), or some less direct interaction involving inter-mediates between the photoreceptor and the porter.

3. Porters and their relation to photoreceptors

If photoreceptors influence transmembrane transport of solutes as a 'primary' event, this must involve catalysed (mediated) transport. The timescale (10^{-6} to 10^2 s) of these primary events is too short for any substantial synthesis or degradation of membrane components. This constraint rules out the modulation of 'lipid solution' transport of solutes through the lipid portion of the membrane as a mechanism of the primary action of photoreceptors. Transport of a neutral solute (such as carbon dioxide or oxygen) by 'lipid solution' can be described by

$$J_{oc} = P(C_o - C_i),$$

where J_{oc} is the net solute flux from phase o to phase i (mol m^{-2} s^{-1}) through a membrane whose permeability coefficient for the solute is P(m s^{-1}) when the concentrations of the solute in phases o and i respectively are C_o and C_i (mol m^{-3}). Since P for a given solute is determined, at a given temperature, by the lipid composition of the membrane, short-term effects of light on J_{oc} must reflect changes in C_0 or C_1, or both, and cannot be construed as reflecting a *membrane* effect on the flux J_{oc}, although the changes in C_0 or C_1 could be products of some other membrane effect of the photoreceptor. An example is light absorption by chlorophyll in the thylakoid membrane, which, by generating NADP and ATP, leads to net carbon dioxide fixation and thus to a net influx of carbon dioxide across the plasmalemma.

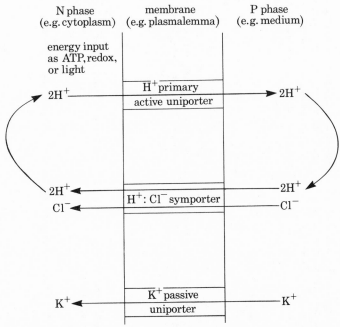

FIGURE 1. Examples of primary active transport (ATP-powered H+ flux from N to P phase) with H+ recirculation coupled to the secondary active transport of Cl− from P to N phase, and electrically driven mediated passive uniport flux of K+ from P to N phase.

Thus primary effects of light on solute transport must be sought in mediated fluxes of solutes, catalysed by intrinsic (transmembrane) protein porters. Figure 1 shows, for a case in which the proton is the working ion, primary active, secondary active, and mediated uniport processes. Light can act as energy source for primary active transport, so that a primary effect of light could be in energizing active transport (the term 'primary event' is taken to be the first detectable obligatory event in the perception–transduction–response sequence, other than the production of the excited state of the chromophore and its return to the ground state). Light could also have informational effects on the activity (i.e. the specific reaction rate of the porter under constant conditions of substrate supply) of any of the three kinds of porter. We note that, as with 'lipid solution' transport, changes in the driving forces acting on the solutes that are being transported (i.e. the chemical activity difference for the transported solute(s) between the two sides of the membrane and, for charged solutes, the electrical potential difference across the membrane) and changes in the free energy available from chemical driving reactions of primary active transport (e.g. increased free energy of ATP hydrolysis under conditions *in vivo*) cannot be construed as primary effects of light.

Table 3 indicates some of the ways in which the photoreceptors mentioned in tables 1 and 2 could influence, by energetic or regulatory means, the rate of solute transport as a primary

TABLE 3. POSSIBILITIES FOR PHOTORECEPTOR–PORTER INTERACTIONS IN A SINGLE MEMBRANE SYSTEM

nature and location of photoreceptor	mechanism of energy or information transfer from photoreceptor to porter	possibility of photoenergization of primary active transport	possibility of direct† photoregulation of primary or secondary active transport, or of mediated passive uniport	references
bacteriorhodopsin, halorhodopsin; integral membrane proteins	in same protein (but may function as trimer) for primary active uniporter; large lateral separation from other porters means direct (protein–protein) interaction is unlikely	yes: of protons by bacteriorhodopsin; chloride ions by halorhodopsin from P to N side of membrane	yes: bacteriorhodopsin and halorhodopsin are laterally segregated from other transporters, so the only direct photoregulation would be of the two rhodopsins themselves	Birge (1981), Schobert & Lanyi (1982)
reaction-centre chlorophyll (P_{700}, P_{680}) proteins (flavoproteins? haem-proteins?); integral membrane proteins	in same protein for primary active transport	yes: of electrons from P to N side of membrane	yes: within molecules, and (?) via conformational change effects on other porters; excitation energy migration out of reaction centres not via antenna pigments	Junge (1977), Raven (1981)
antenna chlorophyll, carotenoid pigment-protein complexes; integral membrane proteins	mostly not in same protein as photochemically active chlorophylls(P_{700} and P_{680}); excitation energy transfer to reaction centre from antenna pigment-proteins in the reaction centre complexes or light-harvesting complexes	yes: by excitation energy transfer to reaction centres	yes: in 'artefactual' protochlorophyll–Ca^{2+} porter membrane systems	Raven (1981)
antenna phycobilins; peripheral thylakoid phycobilisomes, soluble in intrathylakoid space of Cryptophyceae	not in same protein as P_{700} and P_{680}; excitation energy transfer to reaction centre	yes: by excitation energy transfer to reaction centres	yes, but not demonstrated	Gantt (1981)
any integral photoreceptor complex	result of photoinduced conformational change in photoreceptor: alters porter	unlikely that enough energy could be transferred	yes, but not demonstrated	Lemberg (1975), Giaquinta et al. (1975)
any peripheral photoreceptor complex (e.g. phytochrome)	result of photoinduced conformational change in photoreceptor: alters porter	unlikely that enough energy could be transferred	yes, but not demonstrated	Smith (1975)

† Direct means, in this context, an interaction between an integral protein porter and a photoreceptor protein in or on the same membrane by means of excitation energy transfer or conformational energy transfer (see Lemberg 1975).

action of the photoreceptor. By far the best understood of the various couplings of photon absorption to transmembrane fluxes are the energetic couplings of the bacteriorhodopsin–halorhodopsin and the chlorophyll–bacteriochlorophyll systems (see references in table 3). The analysis of these processes has been aided by the relative ease with which the relevant membranes can be extracted and purified, and the large fraction of the membrane protein that consists of the pigment–protein complexes; these factors greatly facilitate the investigation of both the photophysical and photochemical (and the transmembrane flux) aspects of the primary action of light. The much more 'dilute' nature of membrane-associated phytochrome or of flavoprotein (see table 2 of Raven (1981), and § 4a) means that any primary effect of solute transport is less readily investigated; other experimental problems with these two systems will be noted as the discussion proceeds.

4. QUANTITATIVE CONSTRAINTS ON PHOTORECEPTOR ACTION AT MEMBRANES

The two main groups of constraints with which I shall deal are those related to the density of photoreceptors and porters in (or on) membranes, the stoichiometry between photons absorbed and molecules of solute transported by an associated porter and thus to the relation between incident photons and transmembrane solute flux, and the problems of temporal analysis of the relation between photon absorption and solute transport, which is crucial to determining if the effect on solute transport is a primary action of the photoreceptor. It is important to note that such constraints as the low density of photoreceptors on membranes, or the long time taken to complete a photochemical cycle, which are found with phytochrome for example, and not intrinsic to photoreceptors, but are specialities of some of the informational, as opposed to the energetic, photoreceptors (Birge 1981).

(a) Density and specific reaction rates of porters and photoreceptors

Raven (1981) has attempted to relate the density of photoreceptors in (or on) membranes, by means of the specific absorption coefficients of the photoreceptors, to the rate of photon absorption per unit membrane area at a given incident photon flux density. Table 2 of Raven (1981) shows that the density of pigment molecules in the thylakoid membrane (chlorophylls plus carotenoids plus, where they are present, the phycobilins on the membrane) is some 2–3 μmol m^{-2}, while that of flavoprotein at the plasmalemma is some 2.5 nmol m^{-2}, and that of phytochrome at the plasmalemma is only some 0.33 nmol m^{-2} even if all of the cell phytochrome is associated with the plasmalemma. To illustrate the magnitude of fluxes which could be achieved at various photon flux densities if these photoreceptors were to energize active transport, I shall take the hypothetical case of a stoichiometry of 1 photon absorbed for each 1 proton transported for primary active proton transport. The assumed stoichiometry is mechanistically rather than thermodynamically constrained, in that there is much more energy per mole of photons (some 180 kJ mol^{-1} in the red region of the spectrum, and 260 kJ mol^{-1} in the blue region) than is required to pump one mole of protons from an N to a P phase (typical minimum energy requirements of 20–30 kJ mol^{-1}) (see Raven & Smith 1980).

Raven (1981) computed that, with an incident photon flux density of 500 μmol m^{-2} s^{-1} at the wavelength of maximum absorption by the pigment, the flavoprotein and phytochrome densities mentioned above would give proton fluxes of 4.0–4.5 nmol m^{-2} s^{-1}, the similarity of the two fluxes in the face of the different density per unit membrane area being explained by

31

the compensating differences in specific absorption coefficient (see table 1; also Raven 1981). These values are some four orders of magnitude lower than the proton fluxes at the same incident photon flux density in thylakoid membranes. In all cases, however, I have assumed that photon absorption rate determined the proton flux; in reality this is unlikely to be true at an incident photon flux density of 500 µmol m^{-2} s^{-1}, and the proton flux across the thylakoid membrane is likely to be restricted to some 5 µmol m^{-2} s^{-1} by limitations in the reoxidation rate of reduced plastoquinone (Raven 1980; Raven & Smith 1980). For phytochrome the restrictions on the ratio of photons absorbed to protons moved are likely to be less severe, in that the absence of a 'phytochrome unit' analogous to 'photosynthetic units' means that each photon absorbed by phytochrome is used independently rather than being transferred to reaction centres present at some 1 per 400–500 chlorophylls for each of the two photoreactions; however, if energization by phytochrome involves the full P_r–P_{fr}–P_r cycle, I would envisage a specific reaction rate of some 0.83 protons moved per phytochrome molecule per second (see §4b) rather than the 12.7 s^{-1}, corresponding to a flux of 4.1 nmol m^{-2} s^{-1}. Accordingly, the flux might be reduced to 0.27 nmol protons m^{-2} s^{-1}.

Substantial photobehavioural or photomorphogenetic responses occur at a photon flux density of 1 µmol m^{-2} s^{-1} at the wavelength of maximum absorption of the photoreceptor; here the proton fluxes are reduced to 8.2–8.6 pmol m^{-2} s^{-1} at the plasmalemma for phytochrome and flavoprotein, while the thylakoid proton flux would be some 60 nmol m^{-2} s^{-1}. The context in which all of these fluxes should be viewed is that of the 'background' proton recirculation (cf. figure 1) at the eukaryote plasmalemma of 1 µmol m^{-2} s^{-1}, and an irreduceable minimum of uncatalysed 'leak' plus mediated 'slippage' downhill proton fluxes at biological membranes of some 10–20 nmol m^{-2} s^{-1} (Raven 1980; Raven & Smith 1980; Raven & Beardall 1981, 1982; Richardson *et al.* 1983). It is very unlikely that the cell could distinguish between the light-powered proton fluxes of some 10 pmol m^{-2} s^{-1} at a photon flux density of 1 µmol m^{-2} s^{-1} from the 'background' proton flux of up to 1 µmol m^{-2} s^{-1}: it is very likely that such small photoreceptor fluxes of protons would be lost in the noise associated with the background fluxes. However, it is still possible that photoperception of the informational (as opposed to the energetic) type could result from photons absorbed by a photoreceptor having active transport as their primary action. The ciliate *Stentor coeruleus* uses the pigment–protein complex stentorin (with a hypericin-like *meso*-naphthodianthroquinone chromophore) as photoreceptor for its photophobic response (Wood 1970, 1973, 1976; Walker *et al.* 1979, 1981; Song *et al.* 1980). It is possible that the primary action of this chromoprotein is the energizing of active proton transport into cortical vesicles whose bounding membranes contain the stentorin (Walker *et al.* 1981). This possibility is supported by the large amount of stentorin present in the ciliate: the references cited above suggest that *Stentor coeruleus* contains some 0.2 mol m^{-3} of stentorin (specific absorption coefficient 5×10^3 m^3 mol^{-1} m^{-1}), which permits 10 % of the maximum photophobic response to be exhibited at an incident photon flux density of 0.5 µmol m^{-2} s^{-1} of 'red' light. Thus a photobehaviour photoreceptor acting by energizing primary active transport must be present in cells at substantially higher concentrations (0.2 mol m^{-3}) than the phytochrome (0.1 mmol m^{-3}) or plasmalemma-associated flavoprotein (1 mmol m^{-3}) pigments (cf. Raven 1981). The stentorin concentration in *Stentor coeruleus* is, in fact, at the lower end of the range of concentrations found for photosynthetic pigments in oxygen-evolving organisms (table 1 of Raven *et al.* 1979). It is accordingly likely that photoperception by energizing primary active transport requires as a necessary (but not a sufficient) condition that the cells be almost as densely pigmented as phototrophic cells are. Pigments whose primary functions are the harvesting and transduction of light energy for growth of phototrophs also have an important informational role in regulating behaviour and metabolism (Raven 1981; Stoeckenius & Bogomolni 1982).

Turning to the more plausible mechanism (for photoreceptors present at relatively low concentrations in the cell) of the regulation of active or passive mediated fluxes, the maximum specific reaction rates range from 10^2–10^3 s^{-1} (primary active transport of protons or chloride ions), to 10^4 s^{-1} (secondary active transport) to 10^7–10^8 s^{-1} (mediated passive uniport) (Raven 1980; Raven & Smith 1980). The specific reaction rates for mediated passive uniport of 10^7 s^{-1} (calcium (Reuter 1983)) and 10^8 s^{-1} (sodium (Hille 1970)) are close to the limits imposed by the rate of collision of ions with the uniporter from the experimental ion concentrations used (Lauger 1973). Regulation of such porters could achieve very substantial amplification of the photon-induced change in the photoreceptor: the ultimate would be a single molecule of, for example, phytochrome associated with a passive uniporter that had a specific reaction rate of 0 s^{-1} when phytochrome was in the P_r form but had a specific reaction rate of 10^8 s^{-1} when phytochrone was in the P_{fr} form (assuming that the ion concentration were in excess of 100 mol m^{-3} to prevent collision limitation!). Even a modest uniporter specific reaction rate of 10^3 s^{-1} for P_{fr}-activated uniporters present at the same density as phytochrome (0.33 nmol m^{-2}) would give an ion flux of 330 nmol m^{-2} s^{-1}, which (if the ion were the proton) would be greatly in excess of the minimum leakage plus slippage flux of 10–20 nmol m^{-2} s^{-1} (see above), and not much less than the maximum proton recirculation flux of 1 μmol m^{-2} s^{-1}. With a specific photon absorption rate by phytochrome of 0.025 s^{-1} with an incident photon flux density of 1 μmol m^{-2} s^{-1}, only 40 s would be needed to convert all the P_r to P_{fr} and thus activate all the uniporters (cf. Raven 1981). The magnitude of such mediated uniport (down-hill) influxes is such that they would have large electrical effects (depolarized potential difference; increased conductance), which could be part of the transduction process. More specific effects result from the effects of the net ion fluxes on the intracellular free calcium and proton concentrations (cf. the much lower 'normal' bidirectional calcium fluxes at the plant cell plasmalemma (Macklon & Sim 1981)).

For protons, the net passive influx of 330 nmol m^{-2} s^{-1}, with a 500 nm thick layer of cytosol and a proton buffer capacity of 10 mol proton m^{-3} (pH)$^{-1}$ would give a pH change in the cytosol at 0.066 pH s^{-1}, by no means negligible in terms of the 'permitted' change in cytosol pH of perhaps 1 pH (Smith & Raven 1979). An analogous calculation for calcium would, with a calcium buffer capacity of 10 mmol calcium m^{-3} (pCa)$^{-1}$ m^{-1}, yield the same rate of change as for protons, i.e. 0.066 pCa s^{-1} (cf. Raven 1977). While there are homeostatic mechanisms for pH and pCa (both normally about 7.0) in plant cytosol in addition to 'passive buffering', and (as a more temporally pressing matter) some ionic countermovements must occur in the face of a postulated calcium influx of 330 nmol m^{-2} s^{-1}, which, if unbalanced, would give a cell depolarization rate of 3.3 V s^{-1}, modulation of passive uniport by photoreceptors could yield significant electrical and cytosol activity effects that could act as part of a signal transduction sequence. There is evidence that both proton movements and calcium movements across the plasmalemma are important early (tens of seconds or more) products of activation of phytochrome and cryptochrome in plant cells (Marmé 1977; Raven 1981; Senger 1980), so that modulation of active or passive fluxes of these ions could be a primary action site for these photoreceptors.

31-2

(b) *Temporal analysis of photoreceptor action in relation to membrane processes*

The time course of photoreceptor action on solute fluxes is of clear importance in establishing whether the changed solute fluxes are indeed the primary effect of photoreceptor activity. Ideally, the time course of phototransformation of the pigment would be compared with the time course of the change in solute flux, with the latter (for charged solutes) being measured by microelectrode techniques for transplasmalemma electrical potential difference or conductance, or both, to overcome problems of extracellular diffusion lags in the chemical or radiochemical estimation of fluxes. To illustrate how complex such temporal analyses can be, even under very favourable circumstances, I shall first consider a photoelectrical phenomenon of no known photobehavioural or photomorphogenetic significance, i.e. the effect of high photon flux densities of green light on the electrical potential difference at the plasmalemma of the marine coenocytic algae *Acetabularia mediterranea* and *A. crenata* (Schilde 1968; Gradmann 1978). The elegant analysis by Gradmann (1978) showed that green light inhibits the ATP-driven primary active chloride influx of these cells with a lag of less than 40 μs. Even such a short lag is consistent with the diffusion of a low molecular mass chemical messenger from the photoreceptor to the chloride pump over a distance of up to 200 nm, provided that other reactions in the transduction sequence are essentially instantaneous. We note that the slowest reaction in the *Acetabularia* chloride pump reaction sequence has a rate constant in excess of 10^3 s^{-1} when the pump is working at its maximum rate (Gradmann 1978). Although it is likely that the green light effect on the *Acetabularia* chloride pump does involve light absorption by a pump component, or an adjacent membrane component, the possibility of a 200 nm diffusion path being compatible with a lag of 70 μs between initiation of illumination and the first measurable electrical response is salutary in the context of the precision of location 'plasmalemma' photoreceptors by polarized light and microbeam experiments (Haupt 1982), and of the 500 nm thickness of the cytosol of many vacuolate higher plant cells (Macklon 1975).

The shortest lag is an electrical response that can be related to a photobehavioural event is the 1 ms lag between blue light irradiation and a calcium-dependent transcellular electrical potential difference in the chlorophycean flagellate *Haematococcus pluvialis* (Litvin *et al.* 1978). This electrical effect appears to be an essential part of the organism's phototactic response mechanism (cf. Raven 1981). The 1 ms lag is consistent with a diffusion distance inside the cell of 2.8 μm, i.e. a significant fraction of the cell radius and 14 times the diameter of the flagella axes. We note that the total time from 'light on' to the peak electrical potential difference, i.e. about 5 ms at saturating light, is very short compared with the 5 s half-time for the flavoprotein-sensitized reduction of cytochrome *b* in eukaryote plasmalemma, which *may* be a part of the cryptochrome reaction (Brain *et al.* 1977). Although such a slow cytochrome reduction might be a part of the reaction sequence for the cryptochrome effect on electrical potential difference in *Phaseolus vulgaris* hypocotyls (1–5 s lag time (Hartmann & Schmid 1980)) or *Onoclea sensibilis* protonemata (10 s or more (Racusen & Cooke 1982)), the involvement of the much slower (half time of 35 s) reoxidation of the cytochrome by molecular oxygen (Brain *et al.* 1977) in the reaction sequence is dubious, particularly in view of reports of the absence of an oxygen requirement for the induction (Gressel *et al.* 1973) or even the whole response (Kowallick & Gaffron 1967) of certain cryptochrome effects.

The fastest reported phytochrome effect on electrical potential difference (4.5 s after initiation of the red light treatment in *Avena sativa* coleoptiles (Newman 1981)) is slower than some effects of chloroplast-absorbed light on the plasmalemma electrical properties of *Nitella translucens* (1–2 s (Vredenberg 1969)). Thus signals that have to pass from the thylakoid membrane, through the 'tight' inner plastid envelope membrane and the 'leaky' outer plastid envelope membrane, to the plasmalemma have no longer lag time in inducing electrical effects at the plasmalemma than does phytochrome, which may be associated with the N (cytosol) side of the plasmalemma. However, the delay in electrical response to light absorption by P_r does not necessarily imply a substantial diffusion path for a chemical messenger from photoreceptor to porter, because the time taken for the P_r–P_{fr} conversion is by no means negligible in the context of lags of seconds. Briggs & Fork (1969) showed that the half-time for the P_r–P_{fr} conversion was some 0.6 s (as was that for the P_{fr}–P_r photoconversion). This temporal constraint does not apply to reactions in which there is a direct transfer of excitation energy from P_r to for example a primary active transport porter (cf. the assumption made earlier that energization might require the full P_r–P_{fr}–P_r photocycle with, implicitly, a conformational mechanism of energy transfer to a pump); however, the excitation energy transfer mechanism precludes the demonstration of phytochrome's involvement by the criterion of red–far-red reversibility. It is significant that in a number of cases the effect of red light on electrical phenomena, and not merely their induction, can be reversed by far-red light, suggesting that the ion-transport effects are reversibly regulated by the phytochrome system.

We may conclude that the temporal data available do not rule out a direct interaction of cryptochrome or phytochrome with mediated transport at the plasmalemma, but they do not exclude the involvement of an intermediate between photoreceptor and porter that can diffuse over 1 μm or more. We note that the vertebrate retinal rod, a much-studied photoreceptor system, seems to have guanine nucleotide interconversions as the first part of the perception–transduction–response sequence following light absorption by rhodopsin; calcium and calmodulin are involved later, followed by a decrease in the sodium permeability of the plasmalemma (Miller 1981; Birge 1981). Here it seems (Birge 1981) that speed of response is sacrificed to precision, i.e. to decreasing the likelihood of spurious signals, an interesting contrast to the much more rapid response of the energy-transforming bacteriorhodopsin system (see §5).

Lest it be thought that this discussion of the temporal analysis of the interaction of photoreceptors with transmembrane fluxes has ended on a pessimistic note, i.e. that the time course of the effects of phytochrome or cryptochrome is such that the question of 'direct' or 'indirect' interaction of photoreceptor and porters cannot be resolved, it is important to point out that there are other ways of approaching this problem. The most direct (but most technically demanding) would be the 'classical' extraction and reconstitution approach; here isolated and purified membranes (and, if non-integral, photoreceptors) would be tested for the occurrence of photoreceptor–porter coupling. Such experiments could show what components were needed to obtain this coupling; even more convincing would be reconstitution experiments with purified photoreceptor and porter in liposomes (Racker 1976). However, such experiments cannot be conducted until we know a great deal more about cryptochrome and plasmalemma porters in plants.

5. WHY MEMBRANES?

Finally, we may ask why membranes should be involved in photoperception in plants: is it 'evolutionary inertia' or can some selective advantage be construed in membrane-associated photoperception in extant plants? Tackling the 'evolutionary inertia' question first, a plausible evolutionary speculation has been presented by Carlile (1980; cf. Seliger & McElroy 1965; Presti & Delbrück 1978). Early organisms would have been chemoorganotrophs, growing on abiologically photosynthesized organic compounds. As burgeoning life used these organic compounds, the increasingly thin primeval soup might mean that motile organisms with chemotaxis would be at a selective advantage: the chemoreceptors were probably in the plasma-lemma of these organisms (cf. Lengeler 1982). Eventually the evolution of (biological) photo-synthesis overcame this primeval energy crisis, with membrane-associated mechanisms for light-energy transduction (Raven & Smith 1981). Carlile (1980) points out that the association of the membrane-associated photosynthetic photoreceptor with the pre-existing chemotactic system, with its chemosensory apparatus in the plasmalemma, could have led to a phototactic system, with a selective advantage in terms of optimizing the position of the organism in the aquatic photon flux density gradient. This optimization is construed by Carlile (1980) in photo-synthetic terms, although it is also possible that an involvement of the photoreactivation system (Presti & Delbrück 1978) as a photoreceptor for phototaxis related to the avoidance of high photon flux densities of u.v. could have occurred. At all events, the intervening 2×10^9 or so years would seem to have given ample time for evolutionary change of the location of photo-receptors, particularly if the mechanism did not involve net storage of light energy as chemical energy (note that even bioluminescence, which deals in the production of blue quanta of some 260 kJ mol^{-1} energy content, is not membrane-associated (Hastings 1975)).

Having bought forward evidence that phytochrome and cryptochrome do not operate via net membrane-associated energy storage (§4a), it is worth considering what selective pressures might be involved in keeping a photoreceptor system associated with membranes. Birge (1981) and Miller (1981) have recently discussed the attributes of photosensory systems: it would appear that the maintenance of a high signal:noise ratio and the necessary amplification of the signal in the transduction mechanism militates against extremely rapid responses in photo-sensory systems. We note that the very rapid effect of green light on the *Acetabularia* chloride pump does not involve amplification, or a large signal:noise ratio (Schilde 1968; Gradmann 1978). I have already shown that large amplification factors are needed to get good signal:noise ratios for photosensory proton fluxes: the same is true for photosensory calcium fluxes when there are large net calcium fluxes associated with intracellular $CaCO_3$ precipitation (Raven 1981). Overall, unless 'non-nutrient' solute fluxes are being regulated, there would not seem to be any advantage in using transmembrane fluxes as part of a photosensory mechanism in terms of maximizing the signal:noise ratio.

Granted the intrinsic slowness of some photoreceptor events (e.g. phytochrome phototrans-formations), are there situations in which the rapid response (milliseconds rather than seconds) of some photoreceptors can be of advantage to the organism? A good case (see above) is the rapid (5 ms or less) photoelectric response, related to phototaxis, in *Haematococcus pluvialis* (Litvin *et al.* 1978). An important potential 'use' of phototaxis of motile microorganisms, and of plastids in non-motile plants as well as of leaf photonasty in terrestrial plants, is the avoidance of photoinhibition of photosynthesis (Samuelsson & Richardson 1982; Björkman & Powles 1981; Powles & Björkman 1981; cf. Nultsch *et al.* 1981). In *Oxalis oregana* a very good case has been made out for cryptochrome-mediated photonasty in reducing light interception by leaves, and hence in reducing photoinhibition. When a sunfleck (photon flux density 1600 μmol m^{-2} s^{-1}) replaces the normal diffuse forest-floor visible radiation (4 μmol m^{-2} s^{-1}), leaf folding can be completed in 6 min, in which time very little photoinhibition has occurred (Björkman &

Powles 1981; Powles & Björkman 1981). However, the difference between a lag of milliseconds and one of seconds before the arrival of a sunfleck is translated into detectable leaf movement would not seem to be of great moment in this situation. This is true *a fortiori* of the phototaxis of motile phototrophs. Samuelsson & Richardson (1982) showed that the accumulation of *Amphidinium carterae* in a particular region of a photon flux density gradient could be interpreted in terms of maximizing photosynthesis while preventing photoinhibition, which can set in at $80 \ \mu$mol m^{-2} s^{-1} in this shade-adapted dinoflagellate. However, even with a swimming velocity of 0.5 mm s^{-1}, and a large (for open water (Spence 1981)) vertical attenuation coefficient of visible radiation of 0.5 m^{-1}, it would take over half an hour for a dinophyte in stratified water to swim from a definitely photoinhibitory photon flux density of $80 \ \mu$mol m^{-2} s^{-1} to a non-photoinhibitory $40 \ \mu$mol m^{-2} s^{-1}, again making the difference between millisecond and second lags seem irrelevant. For photomorphogenetic rather than photobehavioural responses an even stronger case can be made out for the irrelevance of reducing lag times to below a second.

In conclusion, it is not easy to see any overwhelming advantage in having photoreceptors in plants associated with membranes: their messages are not generally destined for rapid transmission by action potentials, because plants with rapid responses to environmental changes (e.g. carnivores like *Dionaea*) wisely rely on touch rather than shading (Bentrup 1979), thus avoiding the embarrassment of closing their traps on shadows. The hypothesis that photoreceptor action is membrane-associated has generated many useful ideas and experimental results; however, one cannot help feeling that it is a cruel irony that makes the first measureable effects of photoreceptor activity on membrane activity faster than the plant (to our imperfect perception) 'needs', yet not fast enough to avoid ambiguity as to the interaction (direct or indirect) between photoreceptor and porter.

Dr K. Richardson has contributed vigorous discussion of the material in this article. Professor W. Rüdiger has provided important counsel on the possibility of phytochrome's acting as an antenna (sensitizer) *in vivo*.

REFERENCES

Anderson, J. M., Barrett, J. & Thorne, S. W. 1982 Chlorophyll–protein complexes of photosynthetic eukaryotes and prokaryotes: properties and functional organisation. In *Photosynthesis*, vol. 3 (*Structure and molecular organisation of the photosynthetic apparatus*) (ed. G. Akoyonoglou), pp. 301–315. Glenside, Pennsylvania: Balaban International Scientific Services.

Bentrup, H. W. 1979 Reception and transduction of electrical and mechanical stimuli. In *Encyclopedia of plant physiology (new series)* (ed. W. Haupt & M. Feinberg), vol. 7 (*Physiology of movements*), pp. 42–70. Berlin: Springer-Verlag.

Birge, R. R. 1981 Photophysics of light transduction in rhodopsin and bacteriorhodopsin. *A. Rev. Biophys. Bioengng* **10**, 315–354.

Björkman, O. & Powles, S. B. 1981 Leaf movements in the shade species *Oxalis oregana*. I. Response to light level and light quality. *Yb. Carnegie Instn Wash.* **80**, 59–62.

Brain, R. D., Freeberg, J. A., Weiss, C. V. & Briggs, W. R. 1977 Blue light-induced absorbance changes in membrane fractions from corn and *Neurospora*. *Pl. Physiol.* **59**, 948–952.

Briggs, W. R. & Fork, D. C. 1969 Long-lived intermediates in phytochrome transformations. I. *In vitro* studies. *Pl. Physiol.* **44**, 1081–1089.

Carlile, M. J. 1980 The biological significance and evolution of photosensory systems. In *The blue light syndrome* (ed. H. Senger), pp. 3–4. Berlin: Springer-Verlag.

Caubergs, R., Widell, S., Larsson, C. & De Greef, J. A. 1983 Comparison of two methods for the preparation of a membrane fraction of cauliflower inflorescences containing a blue light reducible *b*-type cytochrome. *Physiologia Pl.* **57**, 291–295.

Gantt, E. 1981 Phycobilisomes. *A. Rev. Pl. Physiol.* **32**, 327–347.

Giaquinta, R. T., Ort, D. R. & Dilley, R. A. 1975 The possible relationship between a membrane conformational change and a photosystem two dependent H+ ion accumulation and ATP synthesis. *Biochemistry, Wash.* **14**, 4392–4396.

Gradmann, D. 1978 Green light (550 nm) inhibits electrogenic Cl⁻ pump in the *Acetabularia* membrane by permeability increase to the carrier ion. *J. Membrane Biol.* **44**, 1–24.

Gressel, J., Bar-Lev, S. & Galun, E. 1975 Blue light induced response in the absence of free oxygen. *Pl. Cell Physiol.* **16**, 367–370.

Halliwell, B. 1981 *Chloroplast metabolism. The structure and function of chloroplasts in green leaf cells.* Oxford: Clarendon Press.

Hartmann, E. & Schmid, K. 1980 Effects of UV and blue light on the biopotential changes in etiolated hypocotyl hooks of dwarf beans. In *The blue light syndrome* (ed. H. Senger), pp. 221–237. Berlin: Springer-Verlag.

Hastings, J. W. 1975 Bioluminescence: from chemical bonds to photons. In *CIBA Foundation Symposium* no. 31, pp. 125–146.

Haupt, W. 1982 Light-mediated movement of chloroplasts. *A. Rev. Pl. Physiol.* **33**, 205–233.

Hauska, G., Gabellini, N., Hurt, E., Krinmer, M. & Lockau, W. 1982 Cytochrome *b/c* complexes with polyprenyl quinol:cytochrome *c* oxidoreductase activity from *Anabaena variabilis* and *Rhodopseudomonas sphaeroides* GA: comparison of preparations from chloroplasts and mitochondria. *Biochem. Soc. Trans.* **10**, 340–341.

Hille, B. 1970 Ionic channels in nerve membranes. *Prog. Biophys. molec. Biol.* **31**, 3–32.

Husain, A., Hutson, K. G., Andrew, P. W. & Rogers, L. J. 1976 Flavodoxin from a red alga. *Biochem. Soc. Trans.* **4**, 488.

Junge, W. 1977 Membrane potentials in photosynthesis. *A. Rev. Pl. Physiol.* **28**, 503–536.

Kowallik, W. & Gaffron, H. 1967 Enhancement of respiration and fermentation in algae by blue light. *Nature, Lond.* **215**, 1038–1040.

Lauger, P. 1973 Ion transport through pores: a rate-theory analysis. *Biochim. biophys. Acta* **311**, 423–441.

Lemberg, M. R. 1975 Conformational changes in hemoproteins of the respiratory chain. *Ann. N.Y. Acad. Sci.* **244**, 72–79.

Lengerler, J. 1982 The biochemistry of chemoperception, signal-transduction and adaptation in bacterial chemotaxis. In *Plasmalemma and tonoplast, their functions in the plant cell* (ed. D. Marmé, E. Marré & R. Hertel), pp. 337–344. Amsterdam: North-Holland.

Litvin, F. F., Sineshchekov, O. A. & Sineshchekov, V. A. 1978 Photoreceptor electric potential in the phototaxis of the alga *Haematococcus pluvialis. Nature, Lond.* **271**, 476–478.

MacColl, R. & Burns, D. S. 1978 Energy transfer studies on cryptophycean biliproteins. *Photochem. Photobiol.* **27**, 343–349.

Macklon, A. E. S. 1975 Cortical cell fluxes and transport to the stele in excised root segments of *Allium cepa* L. I. Potassium, sodium and chloride. *Planta* **122**, 109–130.

Macklon, A. E. S. & Sim, A. 1981 Cortical cell fluxes and transport to the stele in excised root segments of *Allium cepa* L. IV. Calcium as affected by its external concentration. *Planta* **152**, 381–387.

Marmé, D. 1977 Phytochrome: membranes as possible sites of primary action. *A. Rev. Pl. Physiol.* **28**, 173–198.

Meeks, J. C. 1974 Chlorophylls. In *Algal physiology and biochemistry* (ed. W. D. P. Stewart), pp. 161–175. Oxford: Blackwells Scientific.

Miller, W. H. 1981 Calcium and cyclic GMP. *Curr. Top. Membrane Tr.* **15**, 441–445.

Mitchell, P. 1979 Direct chemiosmotic ligand conduction in protonmotive complexes. In *Membrane bioenergetics* (ed. C. P. Lee, G. Schatz & L. Ernster), pp. 361–372. Reading, Massachusetts: Addison-Wesley.

Newman, I. A. 1981 Rapid electrical response of oats to phytochrome show membrane processes unrelated to pelletability. *Pl. Physiol.* **68**, 1494–1499.

Ninnemann, H. & Klemm-Wolfgramm, E. 1980 Blue light-controlled conidiation and absorbance change in *Neurospora* are mediated by nitrate reductase. In *The blue light syndrome* (ed. H. Senger), pp. 238–243. Berlin: Springer-Verlag.

Nobel, P. S. 1974 *An introduction to biophysical plant physiology.* San Francisco: W. H. Freeman & Co.

Nultsch, W., Pfau, J. & Ruffer, U. 1981 Do correlations exist between chromatophore arrangement and photosynthetic activity in seaweeds? *Mar. Biol.* **62**, 111–117.

Poff, K. L., Loomis, W. F. & Butler, W. L. 1974 Isolation and purification of the photoreceptor pigment associated with phototaxis in *Dictyostelium discoideum. J. biol. Chem.* **249**, 2164–2167.

Powles, S. B. & Björkman, O. 1981 Leaf movement in the shade species *Oxalis oregana*. II. Role in protection against injury by intense light. *Yb. Carnegie Instn Wash.* **80**, 63–66.

Presti, D. & Delbrück, M. 1978 Photoreceptors for biosynthesis, energy storage and vision. *Pl. Cell Envir.* **1**, 81–100.

Prezelin, B. B. 1981 Light reactions in photosynthesis. In *Physiological bases of phytoplankton ecology* (ed. T. Platt) *(Can. Bull. Fish. Aquat. Sci.* **210**), 1–43.

Racker, E. 1976 *A new look at mechanisms in bioenergetics.* New York: Academic Press.

Racusen, R. H. & Cooke, T. J. 1981 Electrical changes in the apical cells of the fern gametophyte during irradiation with photomorphogenetically active light. *Pl. Physiol.* **70**, 331–334.

Ragan, C. I. 1976 NADH-ubiquinone oxidoreductase. *Biochim. biophys. Acta* **456**, 249–290.

Raven, J. A. 1977 H+ and Ca2+ in phloem and symplast: relation of relative immobility to the cytoplasmic nature of the transport paths. *New Phytol.* **79**, 465–480.

Raven, J. A. 1980 Nutrient transport in microalgae. *Adv. Microb. Physiol.* **21**, 47–226.

Raven, J. A. 1981 Light quality and solute transport. In *Plants and the daylight spectrum* (ed. H. Smith), pp. 375–390. London: Academic Press.

Raven, J. A. & Beardall, J. 1981 The intrinsic permeability of biological membranes to H+: significance for low rates of energy transformation. *FEMS Microbiol. Lett.* **10**, 1–5.

Raven, J. A. & Beardall, J. 1982 The lower limit of photon fluence rate for phototrophic growth: the significance of 'slippage' reactions. *Pl. Cell Envir.* **5**, 117–124.

Raven, J. A. & Smith, F. A. 1980 The chemiosmotic approach. In *Plant membrane transport: current conceptual issues* (ed. R. M. Spanswick, W. J. Lucas & J. Dainty), pp. 161–178. Amsterdam: Elsevier/North-Holland.

Raven, J. A. & Smith, F. A. 1981 H+ transport in the evolution of photosynthesis. *BioSystems* **14**, 95–111.

Raven, J. A., Smith, F. A. & Glidewell, S. M. 1979 Photosynthetic capacities and biological strategies of giant-celled and small-celled macro-algae. *New Phytol.* **83**, 299–309.

Reuter, H. 1983 Calcium channel modulation by neurotransmitters, enzymes and drugs. *Nature, Lond.* **301**, 569–574.

Richardson, K., Beardall, J. & Raven, J. A. 1983 Adaptation of unicellular algae to irradiance: an analysis of strategies. *New Phytol.* **93**, 157–191.

Samuelsson, G. & Richardson, K. 1982 Photoinhibition at low quantum flux densities in a marine coastal dinoflagellate (*Amphidinium carterae*). *Mar. Biol.* **70**, 21–26.

Schilde, C. 1968 Schnelle photoelektrische Effecte der Alge *Acetabularia*. *Z. Naturf.* **23b**, 1369–1376.

Schmid, G. 1980 Conformational changes caused by blue light. In *The blue light syndrome* (ed. H. Senger), pp. 198–204. Berlin: Springer-Verlag.

Schobert, B. & Lanyi, J. K. 1982 Halorhodopsin is a light-driven chloride pump. *J. biol. Chem.* **257**, 10303–10313.

Seliger, H. H. & McElroy, W. D. 1965 *Light: physical and biological action.* New York: Academic Press.

Senger, H. 1980 (ed.) *The blue light syndrome.* Berlin: Springer-Verlag.

Shropshire, W. R. Jr 1980 Carotenoids as primary photoreceptors in blue-light responses. In *The blue light syndrome* (ed. H. Senger), pp. 172–186. Berlin: Springer-Verlag.

Singer, S. J. 1974 The molecular organisation of membranes. *A. Rev. Biochem.* **43**, 805–834.

Smith, F. A. & Raven, J. A. 1979 Intracellular pH and its regulation. *A. Rev. Pl. Physiol.* **30**, 289–311.

Smith, H. 1975 *Phytochrome and photomorphogenesis.* London: McGraw-Hill.

Song, P. S. 1980 Spectroscopic and photochemical characterisation of flavoproteins and carotenoproteins as blue light photoreceptors. In *The blue light syndrome* (ed. H. Senger), pp. 157–171. Berlin: Springer-Verlag.

Song, P. S., Hader, D.-P. & Poff, K. L. 1980 Step-up photophobic response in the ciliate, *Stentor coeruleus*. *Arch. Microbiol.* **126**, 181–186.

Spence, D. H. N. 1981 Light quality and plant response underwater. In *Plants and the daylight spectrum* (ed. H. Smith), pp. 245–275. London: Academic Press.

Stoeckenius, W. & Bogomolni, R. A. 1982 Bacteriorhodopsin and related pigments of Halobacteria. *A. Rev. Biochem.* **51**, 587–616.

Tolbert, N. E. 1981 Metabolic pathways in peroxisomes and glyoxysomes. *A. Rev. Biochem.* **50**, 133–157.

Trebst, A. 1974 Energy conservation in photosynthetic electron transport of chloroplasts. *A. Rev. Pl. Physiol.* **25**, 423–458.

Vredenberg, W. J. 1969 Light-induced changes in membrane potential of algal cells associated with photosynthetic electron transport. *Biochem. biophys. Res. Commun.* **37**, 785–792.

Walker, E. B., Lee, T. Y. & Song, P.-S. 1979 Spectroscopic characterisation of the *Stentor* photoreceptor. *Biochim. biophys. Acta* **587**, 129–144.

Walker, E. B., Yoon, M. & Song, P.-S. 1981 The pH dependence of photosensory responses in *Stentor coeruleus* and model system. *Biochim. biophys. Acta* **634**, 289–308.

Wikstrom, M., Krab, K. & Saraste, M. 1981 Proton-translocating cytochrome complexes. *A. Rev. Biochem.* **50**, 623–655.

Wood, D. C. 1970 Electrophysiological studies of the protozoan, *Stentor coeruleus*. *J. Neurobiol.* **1**, 367–377.

Wood, D. C. 1973 Stimulus specific habituation in a protozoan. *Physiol. Behav.* **11**, 394–354.

Wood, D. C. 1976 Action spectrum and electrophysiological responses correlated with the photophobic response of *Stentor coeruleus*. *Photochem. Photobiol.* **24**, 261–266.

Yakushiji, E. 1971 Cytochromes: algal. *Methods Enzymol.* **23**, 364–368.

Phil. Trans. R. Soc. Lond. B **303**, 419–431 (1983)
Printed in Great Britain

Photoregulation of chloroplast development: transcriptional, translational and post-translational controls?

By G. I. Jenkins, M. R. Hartley and J. Bennett

Department of Biological Sciences, University of Warwick, Coventry CV4 7AL, U.K.

Chloroplast development involves the nucleus, the cytoplasm and the chloroplast of plant cells. This may be illustrated by reference to the two most abundant proteins of the chloroplast: (i) the soluble CO_2-fixing enzyme ribulose 1,5-bisphosphate carboxylase–oxygenase, whose large subunit (LSU) is encoded in chloroplast DNA and synthesized on chloroplast ribosomes and whose small subunit (SSU) is encoded in nuclear DNA, synthesized on cytoplasmic ribosomes in precursor form and transported into chloroplasts, and (ii) the thylakoid-bound light-harvesting chlorophyll *a*/*b* complex, whose pigment components are synthesized in the chloroplast and whose apoproteins resemble the SSU in site of coding and site of synthesis. We have examined the extent to which biosynthetic events in the nucleocytoplasmic compartments are coordinated with those inside the chloroplast during the de-etiolation of pea seedlings. We have examined the levels of LSU, SSU and the light-harvesting chlorophyll *a*/*b* protein (LHCP) by using a highly specific radioimmune assay. The steady-state levels of the corresponding mRNAs have been determined using specific cloned DNA probes. With the SSU, the mRNA and protein levels are near the limit of detection in dark-grown plants but increase markedly under continuous white light, with a lag of about 24 h. The protein appears to be under simple phytochrome control at the level of the steady-state concentration of its mRNA. The LSU also appears to be regulated through the steady-state concentration of its mRNA but in this case the mRNA is not under simple phytochrome control. The LHCP mRNA is readily detectable in dark-grown plants and accumulates further under illumination in a phytochrome-mediated manner. However, the LHCP itself (like chlorophyll) is not detectable in dark-grown plants and accumulates to high levels only under continuous illumination, with a lag of about 6 h. Post-translational control is particularly important in the accumulation of the LHCP: continuous chlorophyll synthesis is required for the stabilization of the protein within the thylakoid membrane, at least during the early stages of chloroplast development.

Introduction

In most angiosperms, chloroplast development goes to completion only in the light (Thomson & Whatley 1980). Two photoreceptors have been implicated unambiguously in this pheno-menon. They are phytochrome (Mohr 1977), which exists in two forms, P_r and P_{fr}, absorbing principally red and far-red light respectively, and protochlorophyllide (Boardman *et al.* 1978) which absorbs principally in the blue and red regions of the spectrum. There is also evidence for a third photoreceptor that is responsive to blue light (Senger 1982).

Protochlorophyllide regulates the synthesis of chlorophyll and the accumulation of chlorophyll-binding proteins. Like all the other steps in the biosynthesis of chlorophyll, the reduction of protochlorophyllide to chlorophyllide by the NADPH-linked enzyme protochlorophyllide reductase (PCR) occurs in the chloroplast. In most angiosperms, but not in all (Adamson &

Hiller 1981), this enzyme is light-dependent. The conversion of the ternary complex NADPH–PCR–protochlorophyllide to NADP–PCR–chlorophyllide requires the absorption of a photon by the protochlorophyllide molecule (Griffiths 1978). No other step in the chlorophyll biosynthetic pathway, including the esterification of chlorophyllide to chlorophyll a and the oxidation of the latter to chlorophyll b, is directly light-dependent. Nevertheless, the light requirement of the PCR reaction means that continuous chlorophyll synthesis requires essentially continuous illumination. Continuous accumulation of certain chlorophyll-binding proteins also requires continuous illumination (Bennett 1983).

TABLE 1. CHLOROPLAST PROTEINS WHOSE mRNAs ARE KNOWN TO BE
REGULATED BY PHYTOCHROME

protein	site of synthesis	effect of red light pulse on mRNA level	references
32 kDa thylakoid protein	chloroplast	increase	1
small subunit of RuBP carboxylase–oxygenase	cytoplasm	increase	2
light-harvesting Chl a/b protein	cytoplasm	increase	2, 3
protochlorophyllide reductase	cytoplasm	decrease	4

References: 1, Link (1982); 2, Tobin (1981a); 3, Apel (1979); 4, Apel (1981).

Phytochrome regulates a wider range of events in chloroplast development than protochlorophyllide but the details of its mechanism of action are less well understood. However, it is clear that phytochrome controls the level of certain mRNAs involved in chloroplast development. The first direct evidence for the regulation of leaf mRNA levels by light was provided by Tobin & Klein (1975), who used *in vitro* cell-free translation systems to show that light increased the levels of the mRNAs for certain unidentified proteins. Subsequently, the two most abundant leaf mRNAs under phytochrome control have been identified as two chloroplast polypeptides: (i) the small subunit (SSU) of the soluble CO_2-fixing enzyme ribulose 1,5-bisphosphate (RuBP) carboxylase–oxygenase (Tobin 1981a) and (ii) the light-harvesting chlorophyll a/b protein (LHCP), the most abundant chlorophyll-binding protein of the thylakoid membrane (Apel & Kloppstech 1978; Apel 1979; Tobin 1981a).

The levels of several mRNAs for chloroplast proteins have now been shown to be controlled by phytochrome (table 1). In each case, the involvement of phytochrome was established by the classical red–far-red reversibility test. Because P_r is converted to P_{fr} by red light and P_{fr} is converted back to P_r by far-red light (Mohr 1977), any physiological response dependent on P_{fr} will, in principle, be elicited by red light but not by far-red light; indeed, far-red light, if administered sufficiently quickly after red light will prevent the expression of the response to red light. With the exception of the mRNA for PCR, a pulse of red light was found to increase the levels of the mRNAs listed in table 1; the level of PCR mRNA was decreased by a pulse of red light.

Three of the mRNAs listed in table 1 (those for LHCP, SSU and PCR) are found in the polyadenylated mRNA fraction and are presumed to be cytoplasmic mRNAs that have been transcribed from nuclear genes. For LHCP mRNA and SSU mRNA, there is direct evidence for the nuclear location of the corresponding genes (Gallagher & Ellis 1982) and for translation of the mRNAs on cytoplasmic ribosomes (Ellis 1981). One of the mRNAs listed in table 1, the mRNA coding for a 32 kDa thylakoid protein, is encoded (Bedbrook *et al.* 1978) and translated (Eaglesham & Ellis 1974) in the chloroplast. Thus, unlike protochlorophyllide, which exerts its effects entirely within the chloroplast, phytochrome regulates aspects of chloroplast development occurring in the nucleus, the cytoplasm and the chloroplast itself.

Another difference between the two photoreceptors is that whereas the effects of the excitation of protochlorophyllide on chlorophyll synthesis persist in darkness only for as long as it takes to convert chlorophyllide to chlorophyll a and chlorophyll b (a matter of minutes), the effects of the formation of P_{fr} from P_r by a brief pulse of red light on the levels of certain mRNAs may persist (and indeed may not be apparent) for many hours. This difference is due

TABLE 2. SUMMARY OF PROCEDURE USED TO ASSAY SINGLE CHLOROPLAST PROTEINS IN
TOTAL LEAF EXTRACTS

(BSA, bovine serum albumin; protein A, immunoglobulin G-binding protein from *Staphylococcus aureus*. For details see Vaessen *et al.* (1981).)

(1) extract leaves with sodium dodecyl sulphate
(2) perform sodium dodecyl sulphate polyacrylamide gel electrophoresis of proteins
(3) transfer proteins to nitrocellulose (NC) by electrophoresis
(4) soak NC with BSA
(5) soak NC with antibody
(6) wash NC
(7) soak NC with ^{125}I-labelled protein A
(8) wash NC
(9) radioautograph

to the stoichiometric involvement of chlorophyllide in chlorophyll synthesis compared with the essentially catalytic role of P_{fr}. One consequence of this difference is that whereas the maximal protochlorophyllide-mediated effects of light on chlorophyll synthesis are seen only under essentially continuous illumination, the phytochrome-mediated effects of light are often nearly maximal under intermittent brief pulses of light separated by long dark periods (e.g. 2 min of white light every 2 h). Indeed, Apel (1979) has shown that the rate of accumulation of translatable polyadenylated LHCP mRNA is the same for several hours whether the barley plants are exposed to continuous illumination with white light or exposed to a single 15 s pulse of red light followed by darkness. However, after about 8 h the level of LHCP mRNA begins to fall in the plants returned to darkness but continues to rise under continuous illumination.

We have addressed ourselves to two major questions that arise from the above results. Firstly, how is the synthesis of the LHCP coordinated with that of chlorophyll a and chlorophyll b, when the apoprotein is synthesized in the nucleocytoplasmic compartment under the control of phytochrome and the pigments are synthesized in the chloroplast under the control of protochlorophyllide? (Note that in posing the question in this way, we are not forgetting that the maximal flux through the chlorophyll biosynthetic pathway under light that is saturating for protochlorophyllide reduction is regulated by phytochrome; see Mohr (1977).)

Secondly, in the synthesis of RuBP carboxylase–oxygenase, how is the synthesis of the SSU, which is encoded in the nucleus and synthesized in the cytoplasm (Ellis 1981), coordinated with that of the large subunit (LSU), which is encoded and synthesized in the chloroplast (Blair & Ellis 1973; Coen *et al.* 1977)? Does phytochrome regulate the synthesis of the LSU? To answer these questions, we have determined the levels of LHCP, LSU and SSU and the levels of their respective mRNAs in pea plants grown under various light regimes. The proteins have been detected by the procedure outlined in table 2, based on the immunological method of

Vaessen *et al.* (1981). The levels of the mRNAs have been determined by hybridization of ^{32}P-labelled cloned DNA probes to leaf mRNA transferred by blotting from agarose gels to nitrocellulose sheets or deposited directly as dots onto nitrocellulose sheets under a slight vacuum.

The accumulation of chloroplast proteins and their messenger RNA molecules during greening

When etiolated pea seedlings 8 days old are transferred to continuous white light (photon fluence rate 100 μmol m^{-2} s^{-1}), chlorophyll accumulation is initially slow but accelerates until, after about 48 h, it is very rapid (figure 1). Whereas chlorophyll *a* is formed from the beginning

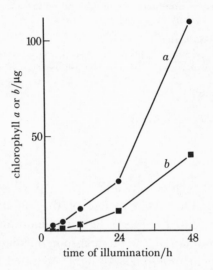

Figure 1. Chlorophyll *a* and chlorophyll *b* contents of apical buds during de-etiolation of pea seedlings 8 days old, under continuous white light. Results expressed per bud.

of the light period, chlorophyll *b* is detectable only after about 6 h. Because most of the chlorophyll *b* and about one-third of the chlorophyll *a* of the photosynthetic membrane occur in the form of the light-harvesting chlorophyll *a/b* complex (Bennett 1983), it would be expected that the accumulation of the apoprotein of this complex, i.e. LHCP, would follow a similar time course to that of chlorophyll *b*. That this is so is shown in figure 2, where the LHCP, LSU and SSU contents of the apical buds of pea seedlings are shown during the first 48 h of de-etiolation. Note that all three polypeptides are at or below the limits of detection in plants grown in the dark for 8 or 10 days. The LHCP becomes detectable after about 6 h of illumination, whereas the SSU becomes detectable only after 24 h. The LSU is also detectable after about 6 h but, unlike the LHCP, the LSU remains at very low levels until the SSU begins to accumulate. These results indicate that the kinetics of accumulation of LHCP, LSU and SSU are different.

What is the mechanism that enables the LHCP to accumulate to comparatively high levels before the subunits of RuBP carboxylase–oxygenase? The prior appearance of the LHCP has already been observed in expanding barley leaves by Viro & Kloppstech (1980). As with other monocots, the expanding leaves of barley display at any one time a gradient of chloroplast development along the blade, with the least mature chloroplasts at the base and the most mature at the tip. The relative abundance of LHCP compared with RuBP carboxylase–oxygenase is higher at the base of the leaf than at the tip. Viro & Kloppstech (1980) have shown that this result can probably be explained in terms of the levels of the corresponding mRNAs. Thus the LHCP mRNA was more abundant than the SSU mRNA near the base of the leaf but less abundant towards the tip. The mRNAs were assayed by *in vitro* translation–immunoprecipitation.

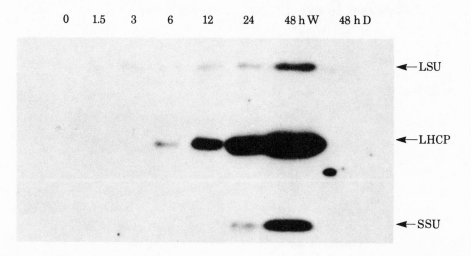

FIGURE 2. Accumulation of LSU, SSU and LHCP during de-etiolation of pea seedlings. Apical buds were harvested after 8 or 10 days of etiolation, or after 8 days of etiolation followed by 1.5–48 h of exposure to white light (W) (100 μmol m^{-2} s^{-1}). Samples corresponding to the sodium dodecyl sulphate extract of one-fiftieth of an apical bud were fractionated by sodium dodecyl sulphate polyacrylamide gel electrophoresis and analysed by the procedure outlined in table 2. The radioautogram shows the position and relative concentrations on the nitrocellulose sheet of the ternary complexes involving LSU (or SSU or LHCP), antibody and ^{125}I-labelled protein A.

We have observed an equivalent *temporal* change in the relative abundance of LHCP mRNA and SSU mRNA during the de-etiolation of pea seedlings. Figure 3 shows the results obtained when total RNA (20 μg) from the apical buds of pea seedlings was fractionated by agarose gel electrophoresis, transferred by blotting to sheets of nitrocellulose and then hybridized with ^{32}P-labelled copy DNA (cDNA) probes complementary to SSU mRNA and LHCP mRNA. Each probe became hybridized to a single region of the filter where the corresponding mRNA was located. The level of SSU mRNA increases steadily from an initial value (in etiolated plants) that is at or below the level of detection to a very high level after 6 days of illumination. Control plants maintained in darkness fail to accumulate detectable levels of the mRNA over the same period. In contrast, the LHCP mRNA is readily detectable in dark-grown plants and accumulates further on illumination.

The results in figure 3 provide a direct visual comparison between the proportional abundances of SSU mRNA and LHCP mRNA in different RNA samples. It is clear from these results that the relative abundance of LHCP mRNA remains relatively constant between 2 and 6 days after the start of illumination, whereas the relative abundance of SSU mRNA increases steadily during this period. Although these data establish that the ratio of SSU mRNA to LHCP mRNA increases during de-etiolation, they should not be interpreted to indicate that LHCP mRNA accumulation is restricted to the first 2 days of illumination. The data in figure 3 underestimate the *absolute* increase in LHCP mRNA and SSU mRNA during de-etiolation because the total RNA content of the leaves begins to increase between 24 and 48 h after the start of illumination (data not shown).

The presence of LHCP mRNA and the absence of SSU mRNA in dark-grown plants together provide an explanation for the fact that the LHCP appears more rapidly than SSU during

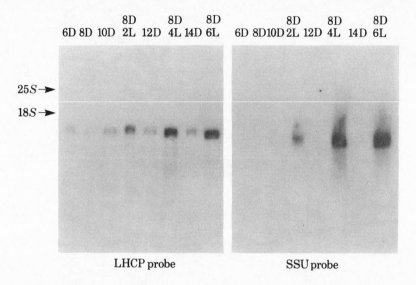

LHCP probe   SSU probe

FIGURE 3. Light-dependent accumulation of SSU mRNA and light-stimulated accumulation of LHCP mRNA during de-etiolation of pea seedlings. Peas were grown in the dark (D) for the times indicated (in days), or transferred into continuous illumination with white light (L) after 8 days. Total RNA was extracted from the apical buds and 20 μg of each sample was fractionated in agarose-formamide gels. The RNA was blotted onto nitrocellulose sheets and hybridized with ^{32}P-labelled nick-translated SSU plasmid (pSSU 160) or LHCP plasmid (pFab 31) (radioactive count of each was 10^7 min^{-1}). See Gallagher & Ellis (1982) for a description of the origins of the clones. The nitrocellulose sheets were washed, dried and radioautographed.

de-etiolation (figure 2). However, the presence of LHCP mRNA in etiolated peas warrants some comment. Although it confirms the result obtained by Cuming & Bennett (1981) when LHCP mRNA was assayed by *in vitro* translation of polyadenylated mRNA followed by immunoprecipitation, it runs contrary to the results of Apel & Kloppstech (1978) and Apel (1979), who could not detect LHCP mRNA in barley leaves by translation *in vitro* and immunoprecpitation. This difference should alert us to the possibility that different plants may show somewhat different patterns of photoregulation of gene expression.

A CRITICAL ASSESSMENT OF MESSENGER RNA ASSAYS

In the assay of mRNA levels, RNA blots such as those shown in figure 3 are superior to *in vitro* translation–immunoprecipitation assays in several respects. Firstly, RNA blots are performed on total RNA and permit the detection of total mRNA levels, whereas translation assays have in the past usually been performed only on polyadenylated mRNA. Secondly, RNA blots provide a built-in check on the degree of mRNA breakdown in the samples. Thirdly, RNA blots do not suffer from the criticism sometimes made of translation assays (but never to our knowledge substantiated for plant mRNA) that translation assays will not detect mRNA sequences that have been chemically modified *in vivo* to render them untranslatable. However, RNA blots have several disadvantages: firstly, it is tedious to perform replicated assays on large numbers of samples; secondly, transfer from agarose gel to nitrocellulose sheet is rarely quantitative; and thirdly, in our hands at least, small to moderate differences in the levels of a mRNA between samples cannot be measured reliably. RNA blots are best suited to the study of large differences in mRNA levels and are the best means of determining whether a DNA probe is hybridizing to a single RNA species.

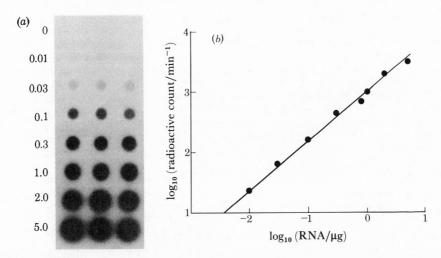

FIGURE 4. Assay of specific mRNAs by dot blotting. Total leaf RNA from plants exposed to 48 h of continuous white light was applied in the indicated quantities in 2.25 M NaCl, 0.225 M trisodium citrate, pH 7.0, to a nitrocellulose sheet held under a slight vacuum in a Perspex manifold. The LHCP mRNA content of each RNA dot was determined by hybridization of ^{32}P-labelled nick-translated LHCP probe (see description of table 3 for details). (*a*) Radioautogram of ^{32}P-labelled hybrids, showing triplication. Total RNA applied (micrograms) is shown on the left. (*b*) Double-logarithmic plot of ^{32}P label in hybrids as a function of total RNA content loaded per dot.

In view of the disadvantages of *in vitro* translation–immunoprecipitation assays and RNA blots, we have had recourse to RNA dot–blots. In this procedure, total RNA in 2.25 M NaCl, 0.225 M trisodium citrate, pH 7.0, is loaded in triplicate on to a nitrocellulose sheet by means of a manifold that applies a slight vacuum to the underside of the sheet. After the RNA is baked onto the sheet, the latter is immersed in a solution containing ^{32}P-labelled DNA probe. After hybridization and washing, the filter is radioautographed to obtain a visual record on the degree of hybridization of the probe to each dot of RNA. The dots are then cut from the nitrocellulose sheet and counted for ^{32}P by scintillation spectrometry to obtain a numerical record of the degree of DNA–RNA hybridization. Figure 4 shows both types of record obtained with a series of concentrations of an RNA sample extracted from pea seedlings that had been de-etiolated under continuous white light for 48 h. The probe is cDNA complementary to LHCP mRNA. The double-logarithmic plot of the numerical data is linear over the entire concentration range from 0.01 to 5 μg of total RNA per dot. Replication is excellent. Routinely, we load 0.8 μg of RNA per dot in these assays. This ensures that high sensitivity is achieved for samples containing little of the mRNA of interest without the risk of overloading the dot with RNA.

DOES PHYTOCHROME REGULATE THE LEVELS OF mRNAs FOR CHLOROPLAST PROTEINS?

We have used the dot–blot assay to determine whether the light-dependent increases in LHCP mRNA and SSU mRNA during de-etiolation of peas (figure 3) are under phytochrome control (table 3). The test for phytochrome involvement is the classical red–far-red reversibility test referred to above.

TABLE 3. PHOTOREGULATION OF mRNAs FOR SSU AND LHCP

(Pea plants were grown in darkness for 8 days and then (a) transferred to white light 100 μmol m^{-2} s^{-1}) for 48 h, (b) maintained in darkness for 48 h, or (c) irradiated as described with red light (15 μmol m^{-2} s^{-1}) or far-red light (8 μmol m^{-2} s^{-1}), or both, and then returned to darkness for 48 h. Triplicate samples of total RNA (0.8 μg) of apical buds were applied to nitrocellulose sheets in 2.25 M NaCl, 0.225 M trisodium citrate, pH 7.0, under slight vacuum. Filters were baked at 80 °C for 2 h *in vacuo* and hybridized with ^{32}P-labelled cDNA probes for either SSU or LHCP mRNAs at 42 °C for 48 h. After being washed at 60 °C in 15 mM NaCl, 1.5 mM trisodium citrate, pH 7.0, the filters were radioautographed and the individual dots excised for scintillation spectrometry. The results show radioactive count per minute ± standard error of triplicates.)

| | hybridization of cDNA probes | | | |
| | SSU | | LHCP | |
light treatment	^{32}P	percentage	^{32}P	percentage
48 h white	2211 ± 76	100	695 ± 42	100
48 h dark	29 ± 2	1.3	128 ± 5	18
15 min red, 48 h dark	215 ± 4	9.7	403 ± 2	58
15 min far-red, 48 h dark	85 ± 1	3.8	239 ± 16	34
15 min red, 15 min far-red, 48 h dark	78 ± 4	3.5	223 ± 1	32

For SSU mRNA, there is only a very low signal for the RNA extracted from dark-grown peas but the signal for RNA extracted from plants exposed to 15 min of red light followed by 48 h of darkness is much higher (and about 10 % of that for plants exposed to continuous white light for 48 h). In contrast, a 15 min pulse of far-red light, either by itself or immediately after a 15 min pulse of red light, results in relatively little production of SSU mRNA. Thus the level of SSU mRNA in peas shows classical far-red light reversibility of red light induction and is concluded to be under phytochrome control. This agrees with the conclusion reached by Tobin (1981 a) for the SSU mRNA of *Lemna gibba*.

Smith & Ellis (1981) found a substantial difference in the levels of hybridizable LSU mRNA between light-grown and dark-grown peas, and Shinozaki *et al.* (1982) observed a large increase in the level of translatable LSU mRNA during de-etiolation of pea plants. We have found (data not shown) that LSU mRNA is readily detectable in etiolated plants and that its concentration increases markedly in response to continuous white light and to a lesser extent in response to a brief pulse of red or far-red light. As yet we are unable to say whether red or far-red light is the more effective in inducing LSU mRNA. For LSU mRNA of *Sinapis alba*, Link (1982) provides data indicating that far-red light induces a higher level of the mRNA than red light. Considerably more work is required before the photoregulation of LSU mRNA levels can be understood, but there are preliminary indications that at least in white mustard (Link 1982) the steady-state level of this mRNA is not under simple phytochrome control. It is possible that phytochrome is not involved directly in the transcription of the genes for LSU in the chloroplast. Note that the presence of LSU mRNA in etiolated pea accords with the synthesis of LSU by isolated intact etioplasts of pea (Siddell & Ellis 1975).

Table 3 confirms the result presented earlier (figure 3) (see also Cuming & Bennett 1981) to the effect that the level of LHCP mRNA found in etiolated peas is comparatively large (about 18 % of the value found for plants exposed to continuous white light for 48 h). When etiolated peas are exposed to a 15 min pulse of red light and then returned to darkness for 48 h, the level of LHCP mRNA increases substantially above the dark level and approaches that obtained under continuous illumination. The inductive effect of red light is significantly reduced by an immediately subsequent pulse of far-red light, indicating that the steady-state level of LHCP mRNA is also under phytochrome control in pea, albeit different in detail from that observed for SSU mRNA. Phytochrome control of the steady-state LHCP mRNA level has also been reported for barley (Apel 1979) and for *Lemna gibba* (Tobin 1981 a).

DOES PHYTOCHROME REGULATE THE LEVELS OF CHLOROPLAST PROTEINS?

Figure 5 shows the levels of LHCP, LSU and SSU in pea plants exposed to various light régimes. From these results, it is clear that the level of SSU, like that of its mRNA, is under phytochrome control, with classical red–far-red reversibility. Note that the level of SSU generated in response to a single 15 min pulse of red light is lower than that produced in response to continuous white light. Thus the level of SSU in peas appears to be determined primarily by the steady-state level of the corresponding mRNA.

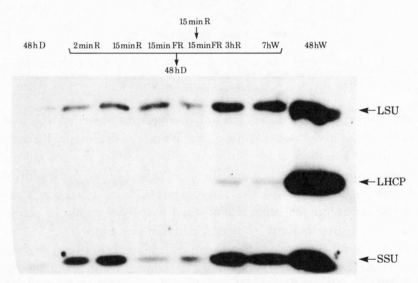

FIGURE 5. Investigation of the involvement of phytochrome in the accumulation of LSU, SSU and LHCP during de-etiolation of pea seedlings. See description of table 3 for a description of plant growth and illumination. See table 2 and figure 2 for details of the radioimmune assay.

The photocontrols on the LSU and LHCP are different from those on the SSU. LSU is induced approximately equally by red and far-red light. This result is in accord with that for the LSU mRNA and indicates that although the level of the LSU is not under simple phytochrome control, it does appear to be determined primarily by the steady-state level of its mRNA, at least in the situation studied here.

The LHCP is also not under simple phytochrome control. Although LHCP mRNA is partly dependent on phytochrome for its accumulation (table 3), it is clear from figures 2 and 5 that

[81]

32-2

LHCP accumulates under continuous illumination but not in dark-grown plants or in plants exposed to light and then returned to darkness, even though in both situations the leaves contain readily detectable levels of LHCP mRNA. There are in fact several situations in which LHCP fails to accumulate in leaves containing the corresponding mRNA. In some cases the LHCP is known to be unstable, and in others instability has been inferred (Bennett 1983). Thus when etiolated pea leaves are de-etiolated for 16 h and then returned to darkness (Bennett 1981), or *Lemna gibba* is subjected in darkness to pulse-chase labelling with L-[^{35}S]methionine (Slovin & Tobin 1982) or bean plants are returned to darkness after a brief period of illumination (Argyroudi-Akoyunoglou *et al.* 1982), LHCP has been shown to break down.

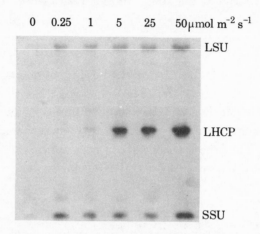

FIGURE 6. Effect of photon fluence rate on the accumulation of LSU, SSU and LHCP during de-etiolation of pea seedlings 8 days old. Plants were exposed to white light of the indicated fluence rates for 48 h before harvest. See table 2 and figure 2 for details of the radioimmune assay.

Bennett (1981) has suggested that LHCP is stabilized against breakdown by some event that occurs during chloroplast maturation. Although the synthesis of chlorophyll *a* and chlorophyll *b* is essential for the stabilization process, it is not sufficient. This point is established by the breakdown of pre-existing LHCP when pea (Bennett 1981), bean (Argyroudi-Akoyunoglou *et al.* 1982) or radish (Lichtenthaler *et al.* 1981) seedlings are placed in darkness. The stabilization of LHCP occurs under continuous illumination (and to a lesser extent in plants exposed to day–night cycles). What irradiance is required for LHCP stabilization?

Our pea plants are routinely grown under warm white fluorescent tubes. The photon fluence rate at the level of the plants is about 50 μmol m^{-2} s^{-1}, compared with about 2000 μmol m^{-2} s^{-1} for full sunlight. When we exposed etiolated pea plants 8 days old to white light of fluence rates in the range 0.25–50 μmol m^{-2} s^{-1} for 2 days and then assayed them for LHCP, LSU and SSU, the results shown in figure 6 were obtained. Dark-grown plants contained little if any of these polypeptides, but plants exposed to even the lowest fluence rate contained both LSU and SSU. Furthermore, the levels of the subunits of RuBC carboxylase–oxygenase were approximately constant over the entire range from 0.25 to 50 μmol m^{-2} s^{-1}. In contrast, only plants exposed to 5–50 μmol m^{-2} s^{-1} contained substantial levels of LHCP. At these photon fluence rates the chloroplasts were clearly able to mature to such a stage that the LHCP could be stabilized. Plants exposed to 1 μmol m^{-2} s^{-1} contained only a trace of LHCP, while plants exposed to 0.25 μmol m^{-2} s^{-1} contained no detectable LHCP. These results provide a new example of a situation where plants contain LHCP mRNA but fail to accumulate the protein.

Concluding remarks

We have studied the photoregulation in pea seedlings of the accumulation of the three most abundant chloroplast polypeptides (namely LSU, SSU and LHCP). Only the SSU appears to be under simple phytochrome control and the data presented above indicate that the level of the polypeptide is determined largely by the level of the corresponding mRNA. The LSU also appears to be regulated largely at the level of the steady-state concentration of its mRNA but in this case the mRNA does not appear to be under simple phytochrome control.

Our present data do not provide any special insight into the mechanisms whereby the levels of the two RuBP carboxylase–oxygenase subunits are coordinated. Although tight coordination has been reported for the LSU and SSU during the cell cycle of *Chlamydomonas reinhardtii* (Iwanij *et al.* 1975), the coordination appears to be rather looser in soybean (Barraclough & Ellis 1979). Our protein assays indicate that in greening pea seedlings, the LSU and SSU accumulate approximately in parallel, but the photocontrols on the two mRNAs are not identical. Thus, for example, LSU mRNA is readily detectable in the dark whereas that for the SSU is not. Furthermore, both the SSU and its mRNA are more abundant after a pulse of red light than after a pulse of far-red light; in contrast, for both the LSU and its mRNA, there appears to be little difference in the inductive abilities of red and far-red light.

The LHCP mRNA is under phytochrome control in pea, as in *Lemna* (Tobin 1981*a*) and barley (Apel 1979). However, this mRNA is present in dark-grown plants to a considerable extent (as much as 18 % of the mRNA found in continuously illuminated plants). This is in marked contrast with barley, where LHCP mRNA is undetectable in etiolated plants (Apel 1979), and with *Lemna*, where LHCP mRNA is very rapidly degraded when green fronds are placed in darkness (Tobin 1981*b*). Clearly, the photocontrols on a given mRNA may differ among plant species.

The absence of the LHCP from dark-grown peas, which nevertheless contain LHCP mRNA, could in principle be due to translational or post-translational controls. Giles *et al.* (1977) and Slovin & Tobin (1982) have provided evidence consistent with translational controls on LHCP mRNA. The 32 kDa translation product detected by Giles *et al.* (1977) is probably the precursor of the LHCP. However, Bennett (1981), Cuming & Bennett (1981) and other authors have argued in favour of post-translational control through the stabilization of LHCP by continuous chlorophyll synthesis. According to the latter view, the accumulation of LHCP is under dual photocontrol via both phytochrome and protochlorophyllide. Further work will be necessary to resolve this question.

This research was supported by grants from the Science and Engineering Research Council to M.R.H. and J.B. We are grateful to Dr S. M. Smith for the generous gift of the plasmids pSSU 160 and pFab 31.

References

Adamson H. & Hiller, R. G. 1981 Chlorophyll synthesis in the dark in angiosperms. In *Photosynthesis, vol. 5 (Chloroplast development)* (ed. G. Akoyunoglou), pp. 213–221. Philadelphia, Pa.: Balaban International Science Services.

Apel, K. 1979 Phytochrome-induced appearance of mRNA activity for the apo-protein of the light-harvesting chlorophyll *a/b* protein of barley (*Hordeum vulgare*). *Eur. J. Biochem.* **97**, 183–188.

Apel, K. 1981 The protochlorophyllide holochrome of barley (*Hordeum vulgare* L.). Phytochrome-induced decrease of translatable mRNA coding for the NADPH:protochlorophyllide oxidoreductase. *Eur. J. Biochem.* **120**, 89–93.

Apel, K. & Kloppstech, K. 1978 The plastid membrane of barley. Light-induced appearance of mRNA coding for the apo-protein of the light-harvesting chlorophyll a/b binding protein. *Eur. J. Biochem.* **85**, 581–588.

Argyroudi-Akoyunoglou, J. H., Akoyunoglou, A., Kalosakes, K. & Akoyunoglou, G. 1982 Reorganization of the photosystem II unit in developing thylakoids of higher plants after transfer to darkness. *Pl. Physiol.* **70**, 1242–1248.

Barraclough, R. & Ellis, R. J. 1979 The biosynthesis of ribulose bisphosphate carboxylase. Uncoupling of the synthesis of the large and small subunits in isolated soybean leaf cells. *Eur. J. Biochem.* **94**, 165–177.

Bedbrook, J. R., Link, G., Coen, D. M., Bogorad, L. & Rich, A. 1978 Maize plastid gene expressed during photoregulated development. *Proc. natn. Acad. Sci. U.S.A.* **75**, 3060–3064.

Bennett, J. 1981 Biosynthesis of the light-harvesting chlorophyll a/b protein. Polypeptide turnover in darkness. *Eur. J. Biochem.* **118**, 61–70.

Bennett, J. 1983 Regulation of photosynthesis by reversible phosphorylation of the light-harvesting chlorophyll a/b protein. *Biochem. J.* **212**, 1–13.

Blair, G. E. & Ellis, R. J. 1973 Protein synthesis in chloroplasts. I. Light-driven synthesis of the large subunit of fraction I protein in isolated pea chloroplasts. *Biochim. biophys. Acta* **319**, 223–234.

Boardman, N. K., Anderson, J. M. & Goodchild, D. J. 1978 Chlorophyll–protein complexes and structures of mature and developing chloroplasts. *Curr. Top. Bioenerget.* B **8**, 35–109.

Coen, D. M., Bedbrook, J. R., Bogorad, L. & Rich, A. 1977 Maize chloroplast DNA fragment encodes the large subunit of ribulose bisphosphate carboxylase. *Proc. natn. Acad. Sci. U.S.A.* **74**, 5487–5491.

Cuming, A. C. & Bennett, J. 1981 Biosynthesis of the light-harvesting chlorophyll a/b protein. Control of messenger RNA activity by light. *Eur. J. Biochem.* **118**, 71–80.

Eaglesham, A. R. J. & Ellis, R. J. 1974 Protein synthesis by chloroplasts. II. Light-driven synthesis of membrane proteins in isolated pea chloroplasts. *Biochim. biophys. Acta* **335**, 396–407.

Ellis, R. J. 1981 Chloroplast proteins: synthesis, transport and assembly. *A. Rev. Pl. Physiol.* **32**, 111–137.

Gallagher, T. F. & Ellis, R. J. 1982 Light-stimulated transcription of genes for two chloroplast polypeptides in isolated pea leaf nuclei. *EMBO Jl* **1**, 1493–1498.

Giles, A. B., Grierson, D. & Smith, H. 1977 *In vitro* translation of messenger RNA from developing bean leaves. Evidence for the existence of stored messenger RNA and its light-induced mobilization into polysomes. *Planta* **136**, 31–36.

Griffiths, W. T. 1978 Reconstitution of chlorophyllide formation by isolated etioplast membranes. *Biochem J.* **174**, 681–692.

Iwanij, V., Chua, N.-H. & Siekevitz, P. 1975 Synthesis and turnover of ribulose bisphosphate carboxylase and of its subunits during the cell cycle of *Chlamydomonas reinhardtii*. *J. Cell Biol.* **64**, 572–585.

Lichtenthaler, H. K., Burkard, G., Kuhn, G. & Prenzel, U. 1981 Light-induced accumulation and stability of chlorophylls and chlorophyll-proteins during chloroplast development in radish seedlings. *Z. Naturf.* **36c**, 421–430.

Link, G. 1982 Phytochrome control of plastid mRNA in mustard (*Sinapis alba*). *Planta* **154**, 81–86.

Mohr, H. 1977 Phytochrome and chloroplast development. *Endeavour* (n.s.) **1**, 107–114.

Senger, H. 1982 The effect of blue-light on plants and microorganisms. *Photochem. Photobiol.* **35**, 911–920.

Shinozaki, K., Sasaki, Y., Sakihama, T. & Kamikubo, T. 1982 Coordinate light-induction of two mRNAs, encoded in nuclei and chloroplasts, of ribulose 1,5-bisphosphate carboxylase/oxygenase. *FEBS Lett.* **144**, 73–76.

Siddell, S. G. & Ellis, R. J. 1975 Protein synthesis in chloroplasts. VI. Characteristics and products of protein synthesis *in vitro* in etioplasts and developing chloroplasts from pea leaves. *Biochem. J.* **146**, 675–685.

Slovin, J. P. & Tobin, E. M. 1982 Synthesis and turnover of the light-harvesting chlorophyll a/b-protein in *Lemna gibba* grown with intermittent red light: possible translational control. *Planta* **154**, 465–472.

Smith, S. M. & Ellis, R. J. 1981 Light-stimulated accumulation of transcripts of nuclear and chloroplast genes for ribulose biphosphate carboxylase. *J. molec. appl. Genet.* **1**, 127–137.

Thomson, W. W. & Whatley, J. M. 1980 Development of non-green plastids. *A. Rev. Pl Physiol.* **31**, 375–394.

Tobin, E. M. 1981a Phytochrome-mediated regulation of messenger RNAs for the small subunit of ribulose 1,5-bisphosphate carboxylase and the light-harvesting chlorophyll a/b-protein in *Lemna gibba*. *Pl. molec. Biol.* **1**, 35–51.

Tobin, E. M. 1981b White light effects on the mRNA for the light-harvesting chlorophyll a/b-protein in *Lemna gibbs* L. G-3. *Pl. Physiol.* **67**, 1078–1083.

Tobin, E. M. & Klein, A. O. 1975 Isolation and translation of plant messenger RNA. *Pl. Physiol.* **56**, 88–94.

Vaessen, R. T. M. J., Kreike, J. & Groot, G. S. P. 1981 Protein transfer to nitrocellulose filters. A simple method for quantitation of single proteins in complex mixtures. *FEBS Lett.* **124**, 193–196.

Viro, M. & Kloppstech, K. 1980 Differential expression of the genes for ribulose 1,5-bisphosphate carboxylase and light-harvesting chlorophyll a/b protein in the developing barley leaf. *Planta* **150**, 41–45.

Discussion

J. W. BRADBEER (*Department of Plant Sciences, King's College London, U.K.*). It is important to recognize that the morphological complexity of the greening pea shoot causes some difficulty in understanding chloroplast development in this system. Leaf development in etiolated peas is arrested at an early stage and the etiolated pea shoot consists largely of stem tissue. On illumination, leaves develop from small leaf initials so that the long lag periods found for the increases in the mRNAs coding for chloroplast polypeptides are not unexpected.

J. BENNETT. The immaturity of etiolated pea buds is of course well known and is one of the attractions of pea as a tissue for studying photoregulation of plant growth. The de-etiolation of peas involves a combination of leaf development and chloroplast development. In the first 40 h of de-etiolation, the total nucleic acid level of pea buds increases approximately threefold while the level of SSU mRNA increases about 50-fold. We take care to exclude the vast majority of the stem tissue when we harvest the buds, which at the stage of growth under study (8 days after germination) consist of the third and higher nodes only.

A. W. GALSTON (*Plant Breeding Institute, Cambridge, U.K.*). The authors state that both the small subunit (SSU) of RuBP carboxylase and its messenger RNA increase in quantity after a red light treatment. Do they have any further information about the comparative kinetics of the post-irradiation appearance of these two entities? Does the increased level of messenger precede the increased level of SSU, as we would expect?

J. BENNETT. In etiolated pea plants the level of SSU mRNA measured by dot hybridization is only about 2% of the level found in plants de-etiolated for 48 h. The increase in SSU mRNA level is initially gradual but accelerates strongly after 12–24 h. The SSU itself is barely detectable by our radioimmune procedure in dark-grown plants or in plants exposed to up to 12 h of illumination. In plants exposed to 24 h of illumination, the SSU is readily detectable and increases a further tenfold, approximately, in the next 24 h.

Phil. Trans. R. Soc. Lond. B **303**, 433–441 (1983)
Printed in Great Britain

Photocontrol of enzyme activation and inactivation

By H. J. Newbury

*Department of Plant Biology, University of Birmingham,
P.O. Box 363, Birmingham B15 2TT, U.K.*

During recent years the question of whether phytochrome regulates certain plant enzyme activities by influencing the rates of enzyme synthesis or by post-translational activation mechanisms has been vigorously debated. There is now good evidence that phytochrome can regulate concentrations of specific messenger RNA populations in some situations. However, the capacity to exert control at a transcriptional level in some systems does not necessarily preclude the possibility that regulation could occur at a post-translational level elsewhere, or even within the same cells. Theoretical considerations apart, the evidence for activation of plant enzymes by phytochrome is not generally strong. Some of the enzymes whose activity is known to be modulated by phytochrome have also been shown to possess post-translational control systems or exist in inactive forms so that the molecular possibilities for such modulation do seem to occur. The lack of direct evidence for the control of such processes by phytochrome may well reflect the technical difficulties involved in this sort of investigation.

Introduction

The plant photoreceptor phytochrome is responsible for mediating a very large number of developmental responses to light (Smith 1975). Many of these responses are clearly apparent without the need for specialized measuring equipment because they result in morphological changes within the plant. At the other end of the spectrum lie a group of molecular responses that may be less obvious but are presumably of great importance during plant photomorphogenesis. A large amount of work has been carried out on a variety of enzymes since it has been found that their extractable activities are increased when etiolated seedlings are exposed to a pulse of red light or to continuous far-red light (Smith *et al.* 1976, 1977; Schopfer 1977). It is part of the dogma of developmental biology that such molecular events represent the means by which the genome exerts its control over development.

When considering the possible mechanisms by which phytochrome may regulate the activities of plant enzymes we normally pose the question 'synthesis or activation?' There is now strong evidence to support the 'synthesis' model; that is that this photoreceptor increases the activity *in vivo* of some enzymes by increasing their rates of production *de novo* and hence their concentrations within plant tissue. There is a tendency to ignore the role of degradative processes in such models even though the degradation rate is of equal importance as the synthetic rate in determining the accumulated level of an enzyme. This is largely because of the technical difficulties associated with the measurement of this process. The term 'activation' is used in this context as an indication that increased extractable activity can, alternatively, result from the post-translational modification of enzyme molecules (figure 1). Activation may mean relief from inhibition. However, the original question 'synthesis or activation?' is to some extent misleading because it implies that the two modes of control are mutually exclusive. There is no reason *a priori* why phytochrome could not regulate both the rate of synthesis and the rate of post-translational activation of a specific enzyme within a particular tissue. Indeed one could argue that this would allow a finer modulation of enzyme activity. Hence the strong evidence that demonstrates phytochrome control of the synthesis of some enzymes and other proteins should not blind us to the possibility that this photoreceptor may also regulate enzyme

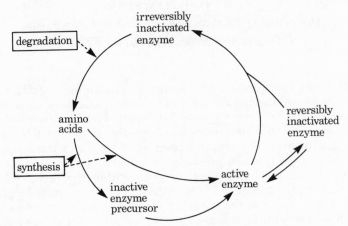

FIGURE 1. Scheme to show possible stages during enzyme turnover.

activities by post-translational mechanisms. However, if the evidence in favour of a role for phytochrome in the activation of plant enzymes is reviewed, the conclusion must be that it is at present very weak. The most obvious explanation for such a conclusion is that this method of control does not actually occur but there are other reasons that must also be considered.

TECHNICAL DIFFICULTIES IN THE DEMONSTRATION OF ACTIVATION

It is currently fashionable to investigate the controls of protein synthesis and gene expression so that perhaps the lack of evidence for phytochrome-regulated post-translational modifications partly reflects the neglect of this subject. There are good academic reasons to explain this neglect. If a research worker decides to investigate the possibility of phytochrome control of enzyme activation there seems to be no single experimental protocol that he can confidently adopt to obtain unequivocal results. Workers who wish to measure changes in the rate of synthesis of an enzyme can do so because they know quite a lot about the various stages of this process, i.e. the transcription of a particular gene followed by the translation of specific messenger RNA molecules. The more this system is investigated the more possible control points are discovered. However, a clear picture of the pathway that produces a specific protein is available and experiments can therefore be designed to measure rates of synthesis.

When, on the other hand, one wishes to test whether phytochrome regulates the post-translational modification of enzyme activities the relative ignorance of these processes is a severe handicap. Strong evidence for such control can only come from (1) an elucidation of the molecular mechanisms by which such enzyme modifications occur, and (2) a demonstration that the rate of modification is regulated by phytochrome. It is really the first of these that has caused problems because information derived from experiments involving eukaryotic organisms shows that there is a plethora of mechanisms by which activation or inactivation can occur.

Conformational changes within an enzyme can modify its activity, and these can occur, for example, as a result of the making and breaking of disulphide bridges. Such a system has been demonstrated in higher plant chloroplasts where light generates reducing power, and transformations in the electron transport chain lead to the reduction of disulphide bonds in several photosynthetically important enzymes (Buchanan 1980). This leads to an activation of these enzymes. In this case phytochrome is not involved and the mechanism depends upon the trapping of light energy by chlorophyll. Enzyme activation has also been shown to occur by the removal of a small polypeptide by proteolytic cleavage (Neurath & Walsh 1976). This has been well documented for certain mammalian enzymes, especially serine proteases, although it is not clear if it occurs in the limited number of these enzymes identified in plants. Allosteric controls offer a further mechanism for the regulation of multimeric enzymes. Several of the enzymes under phytochrome control fall into this category including ascorbate oxidase and phenylalanine ammonia-lyase (PAL). In the latter case allosteric control has been shown to occur upon binding of certain phenolics (Boudet *et al.* 1971). Phytochrome control could therefore theoretically occur by modulation of the cellular concentrations of these compounds. In other cases active enzymes contain a metallic cation, and the presence or absence of this cofactor may represent a control point. Ascorbate oxidase contains copper, which can exist in several states. Both reversible and irreversible inhibitors have been shown to occur in plants, adding further possibilities to the list of post-translational control mechanisms (Wallace 1975). Because it is not possible to predict which of these types of modification may occur in the particular tissue under study it becomes extremely difficult to design satisfactory experiments to investigate enzyme activation.

DIFFERENTIAL DENSITY LABELLING EXPERIMENTS

Some experiments have been designed specifically to distinguish between phytochrome regulation of enzyme synthesis and activation. Some of these involve the use of protein synthesis inhibitors and others use a combination of density labelling and CsCl gradient fractionation. Both methods have their critics and the latter has led to heated debate over the interpretation of results. A typical experimental protocol using differential density labelling as a measure of synthesis begins with seedlings growing in darkness in the presence of water. An extract is made from these seedlings; this is fractionated in a CsCl gradient, and the buoyant density of the band containing the enzyme under study is recorded. In parallel experiments seedlings are transferred to water containing a heavy isotope, commonly deuterium, and some seedlings are exposed to light and some maintained in darkness. After an incubation period extracts are made and again these are fractionated in CsCl gradients and the buoyant densities of the enzyme populations noted. The size of the density shift from the value for control enzyme (that was not exposed to density label) is said to be related to the rate of synthesis of enzyme. Thus if the labelled enzyme population extracted from seedlings exposed to light achieves a higher buoyant density than the labelled population extracted from etiolated seedlings then it is concluded that the enzyme is synthesized at a higher rate in light-treated seedlings. Refinements have been added to the technique but this is the rationale behind differential density labelling.

The technique has been used to compare density shifts of ascorbate oxidase extracted from mustard cotyledons raised in darkness or exposed to far-red light by two independent sets of workers (Attridge 1974; Acton *et al.* 1974). The groups broadly agreed that phytochrome controls the activity of this enzyme by regulating its rate of synthesis. However, after experiments with almost identical methods and tissues the same laboratories could not agree about the role of phytochrome in regulating the activity of PAL, with one group preferring the 'synthesis' (Acton & Schopfer 1975) and the other the 'activation' model (Attridge *et al.* 1974). Interpretation of the results generated by this technique is a complex affair, and the controversy surrounding the PAL studies makes it very difficult for an independent reviewer to reach a clear conclusion on the basis of these results alone.

One weakness of the density labelling and CsCl gradient method as it has been used in the past is that density shifts of only the active forms of enzymes have been followed because gradient fractions have been assayed for catalytic activity. Any inactive form remains un-detected although it may have great physiological significance. Critics of the technique state that results may be impossible to interpret unless it is known whether inactive precursor to the enzyme is either already present or is synthesized as a necessary stage in the production of active enzyme. Lamb & Rubery (1976) have said that differential (comparative) density labelling data of this kind cannot be properly interpreted without an understanding of the mechanism of enzyme 'turnover'. Some of these problems could be overcome if these techniques were to be combined with immunochemical detection of enzyme protein (see later) in gradient fractions rather than relying solely on assayable activity.

Density labelling experiments that employ deuterium oxide have also been criticised on technical grounds since it has been found that in one plant tissue deuterium leads to changes in some cellular membranes. Because the tonoplast is affected, vacuolar proteolytic enzymes are released into the cytoplasm and lead to a large increase in protein degradation rates (Cooke *et al.* 1979, 1980). If this effect is found to be more general then it will affect the interpretation of previous data obtained with mustard seedlings.

Phenylalanine ammonia-lyase in gherkin seedlings

Although differential density labelling experiments have been the subject of much con-troversy in some cases, the results seem only to lend themselves to one reasonable interpretation. This is true of the work carried out on the enzyme PAL in gherkin seedlings. If etiolated gherkin seedlings are exposed to continuous blue light a transient increase in extractable PAL activity occurs. However, after studies of density labelling *in vivo* it was found that active PAL was more heavily labelled if the seedlings were maintained in darkness than if they were illuminated by blue light (Attridge & Smith 1974); these results are most easily explained in terms of PAL activation. The rate of PAL synthesis seems to be lower after the illumination treatment than in darkness. It has been suggested that the higher PAL activity in the light leads to a build-up of reaction products, which inhibit enzyme synthesis (Johnson *et al.* 1975).

In this particular system there is also good evidence for a mechanism for the post-translation modification of PAL, leading to altered activity. A reversible inhibitor of this enzyme has been identified in gherkin hypocotyl extracts (Billett *et al.* 1978). It is non-dialysable, thermolabile, sensitive to proteolytic digestion, hydrophobic and has a molecular mass of less than 20 kDa. The inhibitor preparations will inhibit PAL from a number of plant sources and will also inhibit cinnamic acid 4-hydroxylase. Because this is the enzyme that catalyses the next reaction in the phenylpropanoid pathway and these two enzymes are often regulated coordinately in plants (Smith *et al.* 1967), the inhibitor seems to have a real physiological significance. It does not affect the activity of a wide range of other plant enzymes.

It may be that in this tissue photocontrol of PAL activity is achieved by regulation of the available concentration of inhibitor. To demonstrate this satisfactorily, additional evidence of the significance of the inhibitor *in vivo* is required. Protein synthesis inhibitor studies have already provided an indication that the PAL inhibitor must be continuously synthesized in gherkin seedlings to maintain its effect; treatment of etiolated seedlings with cycloheximide has been shown to lead to a massive increase in PAL activity (Attridge & Smith 1973). A more important feature missing from a proposed activation model in this system is an explanation of the way in which blue light might control the removal of inhibition. A proposal that this illumination may lower the rate of synthesis of the PAL inhibitor seems consistent with the limited experimental evidence but requires much more direct supporting evidence. The previous work on the regulation of PAL activity in gherkin seedlings has thrown up many interesting points and it is to be hoped that these will form the foundation of fresh work in this system.

INACTIVE ENZYMES

A proposal that the photocontrol of enzyme activity occurs, in some cases, by the post-translational regulation of the activity of enzyme molecules contains, by its very nature, a postulation that inactive or partly active forms of enzyme exist *in vivo*. Is there any direct evidence that they do occur? The bulk of the previous work in plant enzymology has quite naturally depended upon the catalytic activity of an enzyme for its detection, so that completely inactive forms of enzyme have remained completely undetectable. Immunochemical techniques do allow a recognition of these proteins when at least some antigenic sites are shared by active and inactive forms. Technically such investigations are time-consuming because of the necessity of raising, purifying and rigorously testing antiserum so that it reacts only with the protein under study.

Such experiments have been carried out on mustard ascorbate oxidase (Newbury & Smith 1981). Antiserum was raised against ascorbate oxidase from a *Cucurbita* species and was tested and purified by using this protein. Extensive tests showed that it cross-reacted with the enzyme from mustard cotyledons, and the simple technique of double-diffusion on agarose plates clearly demonstrated that the enzyme could be detected immunochemically in extracts from unimbibed seeds. These dry seed extracts exhibited no enzymic activity so that the only possible conclusion is that an inactive form of the enzyme does exist. No immunochemical differences could be distinguished between inactive enzyme extracted from seeds and an enzyme population with catalytic activity extracted from cotyledons exposed to light. To quantify the relative concentrations of ascorbate oxidase protein (active plus inactive forms) in extracts from mustard tissues a rocket immunoelectrophoresis method was used. When mustard seedlings were grown in darkness for 72 h the extractable activity of ascorbate oxidase in cotyledons reached a low but easily measurable level. If some seedlings were transferred to continuous far-red light at 36 h the cotyledon extracts showed markedly increased activity of the enzyme (figure 2*a*). When the same extracts were used for the immunochemical assay of ascorbate oxidase protein, the same concentrations were detected in both the etiolated and far-red light treated cotyledons and, indeed, in dry seeds (figure 2*b*). Unlike the results from density labelling experiments carried out on the same system (Attridge 1974; Acton *et al.* 1974) these data seem more consistent with photocontrol by activation of an inactive form.

However, this interpretation makes the assumption that the inactive form of the enzyme found in dry seeds is in fact a precursor to active enzyme *in vivo*; this may not necessarily be so. The inactive enzyme detected here could represent ascorbate oxidase that has previously exhibited activity during seed maturation but has subsequently been inactivated but not degraded. This concept of a time-lag between inactivation and degradation has previously been reported for chalcone synthase, which is detected in an inactivated form in parsley cells some time before it is degraded (Schroder & Schafer 1980). The accumulation of an altered form of β-fructofuranosidase that retained its immunochemical activity but lost its catalytic activity has been

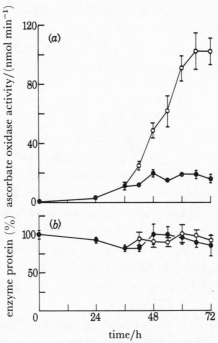

FIGURE 2. Levels of ascorbate oxidase activity and immunochemically detectable oxidase protein in mustard cotyledons. Mustard seedlings were. grown in darkness (●) or transferred to far-red light after 36 h of darkness (○). Samples of cotyledon extract were desalted and assayed for ascorbate oxidase activity (*a*) or concentrated and subjected to rocket immunoelectrophoresis (*b*) to measure the amount of ascorbate oxidase protein, shown as the amount of enzyme protein per cotyledon pair as a percentage of the amount in the unimbibed seed. Standard errors are indicated.

reported in senescent tomato fruit (Nakagawa *et al.* 1980). Furthermore, ascorbate oxidase is reported to be an unusual enzyme in that it can be inactivated because of its catalytic action (Steinman & Dawson 1942). This 'reaction inactivation' is not caused by dehydroascorbate, the reaction product, but appears to be due to a highly reactive form of oxygen produced during the transformation. It seems possible that this process could be involved in the accumulation of immunochemically detectable inactive enzyme. At present the immunochemical data about this enzyme in mustard seedlings cannot be properly interpreted in the absence of information about the physiological significance of this inactive form. Once again a thorough understanding of the 'life-cycle' of an enzyme is lacking and work is in progress to clarify this situation.

THE SELECTIVE ADVANTAGE OF ENZYME ACTIVATION SYSTEMS

It is probably true to say that the cellular concentration of all plant enzymes is subject to regulation at the level of synthesis but that many enzymes have also evolved post-translational control systems. The discussion here is necessarily speculative and concentrates on the relative advantages of these two modes of control in various situations where a plant tissue changes some of its enzymic activities. One apparent advantage of an activation mechanism is the speed of formation of active enzyme. Activation may occur as a single step, whereas a dependence on *de novo* synthesis involves a time lag of several hours during which transcription and translation take place. In some developmental responses this may not matter because the plant tissue may have evolved an integrated and programmed control of gene expression that results in active enzyme becoming available at the exact time that it is required. It seems more likely that activation controls would be more effective in situations where the plant needs to respond quickly to an awaited environmental stimulus.

Seed germination is an interesting process in which to test these hypotheses. In this case either inhibition or an environmental dormancy-breaking signal triggers a large range of metabolic responses. There is evidence that some enzymes are present in seeds in inactive forms that are quickly activated early in germination (Shain & Mayer 1974). Germination and seedling establishment are critical stages in the life cycle of higher plants and it seems that the ability to mobilize certain enzymes rapidly by activation mechanisms represents a selective advantage. Other enzymes are synthesized *de novo* in germinating seeds; the most intensely studied is probably α-amylase, which is produced in barley aleurone cells after the release of gibberellic acid by the embryo (Filner & Varner 1967; Higgins *et al.* 1976). However, there is limited evidence for a third group of enzymes of which a cotton seed protease is an example. Here the messenger RNA for this enzyme is synthesized during seed maturation but is stored in the dry seed and only translated to yield the active enzyme during seed germination (Ihle & Dure 1969, 1972). In this system the delay caused by gene transcription during germination has been removed.

Attempts to relate these concepts to our understanding of the photocontrol of enzyme activities prove extremely difficult. In the germination system it is usually possible to relate changes in enzyme activity to physiological processes occurring within the seed. Many of the enzymes studied are hydrolytic in function and are involved in the degradation of seed storage reserves. In most of the studies involving phytochrome regulation of enzyme activity, seedlings have been transferred from darkness to red or far-red light; this has triggered an increase in the activity of some enzymes but it is usually very difficult to relate this in any causal way to changes that occur within the seedling. The strict catalytic functions of nitrate reductase, phenylalanine ammonia-lyase and ascorbate oxidase may be known but the developmental significance of their increases in activity during photomorphogenesis is not understood. Because of this, speculations on the relative advantages of 'activation' or 'synthesis' modes of control in these situations are not likely to be very profitable.

CONCLUSIONS

If we use the rigorous criteria outlined earlier that require (1) an understanding of the mechanism of post-translational modification, and (2) a demonstration that phytochrome, or the blue light receptor, controls the rate of these modifications, then there does not seem to be a single example of photocontrol (excluding chlorophyll-mediated mechanisms) of plant enzyme activation. However, this need not be because such mechanisms do not exist. Indeed there are some experimental systems that appear very worthy of further study.

REFERENCES

Acton, G. J., Drumm, H. & Mohr, H. 1974 Control of synthesis *de novo* of ascorbate oxidase in the mustard seedling (*Sinapis alba* L.) by phytochrome. *Planta* **121**, 39–50.

Acton, G. J. & Schopfer, P. 1975 Control over activation or synthesis of phenylalanine ammonia lyase by phytochrome in mustard (*Sinapis alba* L.)? A contribution to eliminate some misconceptions. *Biochim. biophys. Acta* **404**, 231–242.

Attridge, T. H. 1974 Phytochrome-mediated synthesis of ascorbic acid oxidase in mustard cotyledons. *Biochim. biophys. Acta* **362**, 258–265.

Attridge, T. H., Johnson, C. B. & Smith, H. 1974 Density-labelling evidence for the phytochrome-mediated activation of phenylalanine ammonia lyase in mustard cotyledons. *Biochim. biophys. Acta* **343**, 440–451.

Attridge, T. H. & Smith, H. 1973 Evidence for a pool of inactive phenylalanine ammonia-lyase in *Cucumis sativa* seedlings. *Phytochemistry* **12**, 1569–1574.

Attridge, T. H. & Smith, H. 1974 Density-labelling evidence for the blue-light-mediated activation of phenylalanine ammonia-lyase in *Cucumis sativus* seedlings. *Biochim. biophys. Acta* **343**, 452–464.

Billett, E. E., Wallace, W. & Smith, H. 1978 A specific and reversible macromolecular inhibitor of phenylalanine ammonia-lyase and cinnamic acid-4-hydroxylase in gherkins. *Biochim. biophys. Acta* **524**, 219–230.

Boudet, A., Ranjeva, R. & Gadal, P. 1971 Propriétés allostériques specifiques des deux isoenzymes de la phenylalanine-ammoniaque lyase chez *Quercus pedunculata*. *Phytochemistry* **10**, 997–1005.

Buchanan, B. B. 1980 Role of light in the regulation of chloroplast enzymes. *A. Rev. Pl. Physiol.* **31**, 341–374.

Cooke, R. J., Grego, S., Oliver, J. & Davies, D. D. 1979 The effect of deuterium oxide on protein turnover in *Lemna minor*. *Planta* **146**, 229–236.

Cooke, R. J., Grego, S., Roberts, K. & Davies, D. D. 1980 The mechanism of deuterium oxide-induced protein degradation in *Lemna minor*. *Planta* **148**, 374–380.

Filner, P. & Varner, J. E. 1967 A test for de novo synthesis of enzymes: density labelling with H_2O^{18} of barley α-amylase induced by gibberellic acid. *Proc. natn. Acad. Sci. U.S.A.* **58**, 1520–1526.

Higgins, T. J. V., Zwar, J. A. & Jacobsen, J. V. 1976 Gibberellic acid enhances the level of translatable mRNA for α-amylase in barley aleurone layers. *Nature, Lond.* **260**, 166–168.

Ihle, J. N. & Dure, L. S. 1969 Synthesis of a protease in germinating cotton cotyledons catalysed by mRNA synthesised during embryogenesis. *Biochem. biophys. Res. Commun.* **36**, 705–710.

Ihle, J. N. & Dure, L. S. 1972 The developmental biochemistry of cottonseed embryogenesis and germination. III. Regulation of the biosynthesis of enzymes utilised in germination. *J. biol. Chem.* **247**, 5048–5055.

Johnson, C. B., Attridge, T. H. & Smith, H. 1975 Regulation of phenylalanine ammonia-lyase synthesis by cinnamic acid: its implication for the light-mediated regulation of the enzyme. *Biochim. biophys. Acta* **385**, 11–19.

Lamb, C. J. & Rubery, P. H. 1976 Interpretation of the rate of density labelling of enzymes with ^2H_2O. Possible implications for the mode of action of phytochrome. *Biochem. biophys. Acta* **421**, 308–318.

Nakagawa, H., Iki, K., Hirata, M., Ishigawa, S. & Ogura, N. 1980 Inactive β-fructofuranosidase molecules in senescent tomato fruit. *Phytochemistry* **19**, 195–197.

Neurath, H. & Walsh, K. A. 1976 Role of proteolytic enzymes in biological regulation. *Proc. natn. Acad. Sci. U.S.A.* **73**, 3825–3832.

Newbury, H. J. & Smith, H. 1981 Immunochemical evidence for phytochrome regulation of the specific activity of ascorbic oxidase in mustard seedlings. *Eur. J. Biochem.* **117**, 575–580.

Schopfer, P. 1977 Phytochrome control of enzymes. *A. Rev. Pl. Physiol.* **28**, 223–252.

Schroder, J. & Schafer, E. 1980 Radio-iodinated antibodies, a tool in studies on the presence and role of inactive enzyme forms: regulation of chalcone synthase in parsley cell suspension cultures. *Archs Biochem. Biophys.* **203**, 800–808.

Shain, Y. & Mayer, A. M. 1974 Control of seed germination. *A. Rev. Pl. Physiol.* **25**, 167–193.

Smith, H. 1975 *Phytochrome and photomorphogenesis*. London: McGraw-Hill.

Smith, H., Attridge, T. H. & Johnson, C. B. 1976 Photocontrol of enzyme activity. In *Perspectives in experimental biology* (ed. N. Sunderland), pp. 325–336. Oxford: Pergamon.

Smith, H., Billett, E. E. & Giles, A. B. 1977 The photocontrol of gene expression in higher plants. In *Regulation of enzyme synthesis and activity in higher plants* (ed. H. Smith), pp. 93–127. London: Academic Press.

Steinman, H. G. & Dawson, C. R. 1942 On the mechanism of the ascorbic acid – ascorbic acid oxidase reaction. The hydrogen peroxide question. *J. Am. chem. Soc.* **64**, 1212–1219.

Wallace, W. W. 1975 Proteolytic inactivation of enzymes. In *Regulation of enzyme synthesis and activity in higher plants* (ed. H. Smith), pp. 177–195. London: Academic Press.

Discussion

J. W. BRADBEER (*Department of Plant Sciences, King's College London, U.K.*). Dr Newbury only gave a very small mention of the chloroplast carbon pathway enzymes known to be synthesized by dark-grown etiolated leaves and to have their synthesis promoted by phytochrome, although those carbon pathway enzymes that are subject to activation are activated by chloroplast metabolites and not by phytochrome. The state of knowledge with respect to the carbon pathway enzymes has been obtained from the application of a very wide range of techniques.

H. J. NEWBURY. I recognize that the group of enzymes to which Professor Bradbeer refers represent a well characterized example of light-induced post-translational activation in plants; my treatment of this area has been rather cursory not because this is not an interesting and important system for study but because the meeting has mainly been directed towards a discussion of photoreceptors not directly related to the photosynthetic process. Apart from being of general interest to students of plant molecular biology, research into the chloroplast carbon pathway enzymes has provided one of the few examples where the actual mechanism of post-translational modification has been properly elucidated.

Phil. Trans. R. Soc. Lond. B **303**, 443–452 (1983)
Printed in Great Britain

Is P_{fr} the active form of phytochrome?

BY H. SMITH

Department of Botany, University of Leicester, University Road, Leicester LE1 7RH, U.K.

A selective but critical assessment of the published relations between spectrophoto-metric measurements of phytochrome parameters *in vivo* and physiological responses is presented. Although a number of correlations between response and the apparent concentration of P_{fr} (i.e. $[P_{fr}]$) have been reported, these are counterbalanced by authenticated cases of lack of correlation. The reported relations between $[P_{fr}]$ and response are not uniform, nor are they subject to ready interpretation. It seems useful to consider the responses of dark-grown and light-grown plants to be in principle different, with partly de-etiolated plants exhibiting intermediate responses. Dark-grown plants show high sensitivity to small changes in $[P_{fr}]$ at low $[P_{fr}]$, whereas light-grown plants show high sensitivity at high $[P_{fr}]$. Current data are not easily in-terpretable in terms of P_{fr} being the only active component of the phytochrome system.

INTRODUCTION

The concept of P_{fr} as the active form of phytochrome is central to current thought on the cellular mechanism of action of the photoreceptor. It forms the basis of the sophisticated analysis of the molecular properties of phytochrome *in vitro*, which seeks to define differences between P_r and P_{fr} sufficient to account for the presumed biological activity of P_{fr} (see for example Quail, Furuya and Rüdiger, this symposium), and lies at the heart of almost every attempt to in-terpret complex physiological data. Indeed, most papers on phytochrome begin with an assertion similar to: 'It has been generally agreed that P_{fr} is the physiologically active form of phytochrome' (quoted from the initial sentences in Whitelam & Johnson (1981) and Schmidt & Mohr (1982)). The view is so deep-seated that authors in general do not feel it is necessary to provide any supporting evidence, and assertions such as the one quoted above are normally followed by unhelpful references to the text-books written by Mohr (1972) and Smith (1975). 'P_{fr} as the active form', therefore, seems to have passed from the status of a theory to that of a central dogma. My objective in this article is selectively to examine the evidence for this 'central dogma', with particular reference to the action of phytochrome in light-grown plants. First, however, I wish to draw attention to the loose nature of the concept, as it is normally formulated. The statement 'P_{fr} as the active form' is imprecise unless it also means 'P_{fr} as the *only* active form'; in this article, the exclusivity expressed in the latter formulation is assumed to hold.

CORRELATIONS BETWEEN MEASURED PHYTOCHROME LEVELS AND PHYSIOLOGICAL RESPONSES

The concept of P_{fr} as the active form was generated by the earliest experiments on phyto-chrome-mediated responses, which demonstrated that red light induces developmental effects whereas far-red has no effect except after red; this generalization came from physiological responses observed in dark-imbibed seeds, dark-grown seedlings, and in light-grown plants given brief light periods during a photoperiodically inductive dark period (Borthwick *et al.*

1952, 1954; Parker *et al.* 1946, 1949). Because only a small quantity of red light is sufficient to elicit substantial developmental changes, it was natural to assume that the form produced by that light – i.e. P_{fr} – actively induces the changes via some physiological amplification mechanism. The logical alternative, that P_r acts to inhibit the developmental changes, was justifiably ruled out on the grounds that a large proportion of an active inhibitor would need to be removed before significant response would be detected.

Because it has not yet been possible to associate a biological activity with purified phytochrome *in vitro*, the only evidence on whether or not P_{fr} possesses unique biological activity not exhibited by other forms of phytochrome comes from correlations between spectrophotometrically determined phytochrome levels *in vivo* and physiological responses. There are many problems inherent in this limitation in experimental approach; for example:

(*a*) it is only possible to measure the tissue average of phytochrome content;

(*b*) only relative measures of phytochrome can be obtained;

(*c*) the measurements are affected by the optical properties of the tissues, which may change during the treatments given;

(*d*) some parameters of phytochrome are not measurable, e.g. the level of photoconversion intermediates;

(*e*) should the responses be quantitatively correlated to phytochrome parameters determined at the time of light treatment or at the time at which the responses are measured?

With such serious conditions placed upon the experimental approach it is quite surprising that any correlations at all have been observed; on the other hand the existence of so many conditional factors allows wide latitude in interpretation of conflicting data! Perhaps the most important point about correlative evidence, however, is that it can never be anything more than circumstantial. No matter how many instances of positive correlations are accumulated, it only takes one single definitive and authenticated negative correlation to disprove the working hypothesis.

This article is concerned principally with data from light-grown seedlings but some coverage of the correlations observed with dark-grown seedlings is necessary to set the scene.

(*a*) *Dark-grown seedlings*

For dark-grown seedlings given brief light treatments essentially three types of positive correlations have been observed. These are: (1) the response is related to the logarithm of P_{fr} concentration (i.e. $\log[P_{fr}]$); (2) the response shows a biphasic linear relation to $[P_{fr}]$; (3) the response is related to $[P_{fr}]$ on a threshold basis.

Early observations by Hillman and others (Hillman 1965, 1966; Loercher 1966) showed that response to brief treatments of red light or to mixed red and far-red light became saturated at relatively low levels of P_{fr}/P_{total} and that the relation below saturation approximated to a logarithmic curve. Subsequently, Mandoli & Briggs (1981) were able to pull together much of the early data and show that the relations between response and $[P_{fr}]$ could be plotted together on a log–linear basis. Mohr and colleagues (Drumm & Mohr 1974; Steinitz *et al.* 1979), however, studying anthocyanin synthesis in mustard seedlings described the relation between $[P_{fr}]$ and response as biphasic–linear. This is expressed as a much greater degree of sensitivity to changes in $[P_{fr}]$ at low $[P_{fr}]$ than at higher $[P_{fr}]$. The third type of relation was shown by Oelze-Karow & Mohr (1970, 1973) studying lipoxygenase in mustard seedlings, where they found that the increase in activity of lipoxygenase was switched on at a $[P_{fr}]$ below a certain critical, threshold value; there was no graded response at all in this case.

If we assume from these correlations that P_{fr} is the only active component the question arises as to whether P_{fr} acts differently in the three types of relation. A logarithmic relation between the concentration of an effector and the response to that effector is not common in biology, although other examples can be found, e.g. the relation between the concentration of certain plant growth substances and response is in many cases logarithmic. Depending on the reliability of the data, however, it is not always easy to distinguish between a logarithmic relation and a hyperbolic relation, which might be more understandable in terms of classical biochemistry (i.e. Michaelean analysis). The biphasic–linear relation of Drumm & Mohr (1974) may also be a manifestation of a hyperbolic function although the linear fits are impressive. The authors interpret their data in terms of a biphasic cooperativity between P_{fr} molecules associated with a presumed membrane matrix; the problems of understanding this biphasic relation are not trivial. The threshold relation is perhaps the easiest to understand biologically (e.g. some form of cascade control sequence), but it seems to have limited importance judging from the rarity of confirmatory reports.

Even though it might be difficult to reconcile the three different types of relation between $[P_{fr}]$ and response, nevertheless on the evidence quoted above it is not necessary to postulate activity associated with any component other than, or additional to, P_{fr}. There are, however, some quite celebrated examples of a lack of correlation between $[P_{fr}]$ and response. These have generally been termed 'paradoxes' and the most serious is the so-called *Zea* paradox in which red light induces a response that saturates at fluences well below those required for detectable photoconversion of P_r to P_{fr}, and which is reversed by fluences of far-red that establish higher P_{fr}/P_{total} than does the red light fluence required to saturate the response (Chon & Briggs 1966; Briggs & Chon 1966). I know of no acceptable argument for ignoring these data, but they seem generally to be discounted in considerations of whether or not P_{fr} is the only active form of phytochrome. Similar extreme sensitivity to very low levels of red light has been reported more recently in a number of other responses observed with totally dark-grown seedlings (Raven & Shropshire 1975; Small *et al.* 1979; Mandoli & Briggs 1981) and dark-imbibed seeds (VanDerWoude & Toole 1980). In these cases the organisms show substantial responses to $[P_{fr}]$ well below those that can be detected by spectrophotometry *in vivo*. Such responses – which have been termed 'very low fluence responses' by Mandoli & Briggs (1981) – may be quite general with dark-grown plants, having been missed previously by investigators who used green 'safe lights' even for their dark controls; exposure of the material to any light whatsoever is sufficient to induce the very low fluence response. The earlier observations of Hillman and others did, however, show significant physiological response to light treatments that established a $[P_{fr}]$ not detectable by spectrophotometry (see Hillman 1967, 1972 for reviews). Sensitivity to low $[P_{fr}]$ may also be markedly affected by a pretreatment with light. This has been shown particularly with anthocyanin synthesis in mustard cotyledons, where so-called 'sensitivity amplification' is induced by pretreatment for several hours with light, itself apparently operating through phytochrome (Mohr *et al.* 1979; Johnson 1980). Since the long-term light pretreatment leads to partial de-etiolation, this topic is dealt with in more detail below.

When considering the data for dark-imbibed seeds and dark-grown seedlings, therefore, it seems useful to draw attention to the extremely high sensitivity to changes in $[P_{fr}]$ at low levels of $[P_{fr}]$, and to the variety of different relations exhibited between $[P_{fr}]$ and response.

[99]

(b) High-irradiance responses

High-irradiance responses are responses of dark-grown seedlings to continuous light, and many different action spectra have been produced, most showing peaks of action in the blue and in the red or far-red (or both) wavelength bands. The action in the red and far-red band was shown by Hartmann (1966, 1967 a), using the now classical bichromatic irradiation techniques, to be mediated by phytochrome. In some of the best-studied examples – e.g. lettuce hypocotyl growth (Hartmann 1967 b) and mustard hypocotyl growth (Beggs et al. 1980) – the response is maximal in the far-red, where very low $[P_{fr}]$ is established. The action spectra in general are dependent upon the length of irradiation treatment and the pretreatment conditions and also on the level of response chosen as being a standard response (see Beggs et al. 1980; Holmes & Schäfer 1981). During the period of the irradiation the concentration of total phytochrome is constantly changing and thus any possible relation between $[P_{fr}]$ and response must be quite complex. Although it is possible that the responses are related to the integral of $[P_{fr}]$ over the time period of the irradiation treatment, a direct relation between response and $[P_{fr}]$ is not obvious.

On the other hand, even on simple models, under continuous irradiation $[P_{fr}]$ becomes independent of wavelength and strongly dependent on fluence rate (Schäfer & Mohr 1974; Gammerman & Fukshansky 1974). Heim & Schäfer (1982) have shown that both $[P_{fr}]$ and P_{fr}/P_{total} show strong fluence-rate dependence under continuous red light and repeated brief red light treatments, and that these effects on phytochrome parameters at least partly parallel the effects of different fluence rates of red light on hypocotyl growth. Recently, the same authors (Heim & Schäfer 1983) have tackled the closely related question of whether or not the fluence-rate dependence of mustard hypocotyl growth under continuous far-red (Beggs et al. 1980; Holmes & Schäfer 1981) is paralleled by a fluence-rate dependence of $[P_{fr}]$. Their data show a lack of correlation between the measured phytochrome parameters and the observed responses, and they conclude that P_{fr} cannot be solely responsible for the action of phytochrome in the high-irradiance response, although they wisely do not speculate upon which other component is likely also to be involved.

Others have also concluded that P_{fr} cannot be the only active form of phytochrome in the high-irradiance response. Johnson & Tasker (1979) concluded that P_{fr} interacted with a cycling-driven process (see below for detailed treatment). Recently, Bartley & Frankland (1982) have investigated a high-irradiance response in seed germination, in which the response operates antagonistically to that elicited by brief red light; here again it was proposed that phytochrome cycling was responsible for the inhibitory action of continuous light.

(c) Light-grown plants

In recent years, studies of the photomorphogenetic reactions of light-grown plants have become fashionable. The transition from etiolation growth to growth under light takes a substantial time and during the transition period plants appear to exhibit responses that are characteristically different from either those shown by dark-grown plants or those shown by mature, fully de-etiolated plants. It is therefore necessary to establish criteria for the responses that appear to be characteristic of mature, de-etiolated, light-grown plants. For complete de-etiolation it seems necessary to have quite a long period (24–28 h at least) of relatively high fluence rate (more than ca. 50 μmol m^{-2} s^{-1}) of light, which establishes a relatively high P_{fr}/P_{total}.

During the transition period the extension growth rate of the seedling drops from the high rate characteristic of etiolation growth to the very much lower rate characteristic of light-grown seedlings. In mustard, for example, elongation growth in the dark can be of the order of 20–40 µm min^{-1}, whereas in a fully de-etiolated young seedling the growth rate is commonly 1–2 µm min^{-1}; as the seedlings mature in the light, growth rate usually increases again, reaching 5–8 µm min^{-1} after 3–4 weeks. Mustard seedlings growing at the low rate in the light initially retain the same growth rate when transferred to darkness, although within 1 or 2 hours the growth rate begins to decline even further (Child & Smith, unpublished data). A reasonable criterion of de-etiolation would therefore be the establishment of a minimum growth rate. Such seedlings growing under fluorescent white light, when given additional far-red light – which depresses P_{fr}/P_{total} – exhibit a substantial increase in extension growth, and it is this far-red mediated increase in extension growth that is the true characteristic of the photomorphogenetic responses of the fully de-etiolated plant (Holmes & Smith 1975, 1977; Morgan & Smith 1976, 1978). As a rule of thumb, it seems that seedlings that still have extending hypocotyls react in a transitional way to light, and the far-red-induced growth increases do not become fully apparent until hypocotyl elongation has ceased and elongation growth of the seedling is solely due to the growth of true internodes.

If we concentrate first on seedlings given short, partly de-etiolating light treatments, it is now recognized that subsequent phytochrome-mediated phenomena show unusual relations between $[P_{fr}]$ and response. As mentioned briefly above, the partial de-etiolation treatment can lead to a large increase in the subsequent sensitivity of responses to brief red light treatments; this is described as 'sensitivity amplification' and interpreted as an increased response to low $[P_{fr}]$ (Mohr et al. 1979; Johnson 1980). Sensitivity amplification is a relatively short-lived phenomenon, with the increase in response disappearing over about 6 h (Schmidt & Mohr 1982). Schmidt & Mohr (1983) have recently concluded that the mechanism of sensitivity amplification does not reside in an acceleration of the transduction process between phytochrome and response (anthocyanin synthesis), whereas Mohr & Schäfer (this symposium) discuss the possible ways in which amplification of the 'P_{fr} signal' may be achieved. It seems unlikely, however, that an increased level of 'receptor sites' for P_{fr} could explain the observed amplification (Oelmüller & Mohr 1983). Whitelam & Johnson (1981), investigating sensitivity amplification of the phytochrome control of nitrate reductase activity in mustard cotyledons, showed that amplification was a function of the fluence rate of the light pretreatment. They concluded that the process occurring during the pretreatment causes sensitivity amplification through the time-dependent and fluence-rate-dependent (i.e. cycling-dependent) synthesis or transport of a component that interacts with P_{fr}. Thus P_{fr} is considered to be active, but its activity is considered to be increased through interaction with another component whose level is dependent on phytochrome action, operating in a manner *independent* of $[P_{fr}]$.

Fully de-etiolated seedlings exhibit a type of phytochrome-controlled growth response that seems, from the phenomenological point of view, different from both the red–far-red inducible responses and the high-irradiance responses shown by dark-grown plants. Stem extension in a light-grown plant, when measured over a period of days, is linearly related to the P_{fr}/P_{total} calculated from the spectral distribution of the actinic radiation (Morgan & Smith 1976, 1978). There are problems with estimations of P_{fr}/P_{total}, because it is clearly impossible to measure phytochrome spectrophotometrically in green plant tissues, and P_{fr}/P_{total} is calculated from the spectral photon distribution of the incident radiation. Others, however, have reported similar linear relations in light-grown plants (Vince-Prue 1977; Holmes, this symposium). The linear relation between extension growth and P_{fr}/P_{total} has now been seen in a wide range of species, and may be regarded as a characteristic of light-grown plants (Morgan & Smith 1979; Smith 1982).

The responses of light-grown plants to a change in the red:far-red ratio are very rapid. Transducer measurements of growth rate in mustard seedlings have shown a 10–15 min lag between the onset of additional far-red and the first detectable increase in growth rate (Morgan *et al.* 1980). After the 10–15 min lag the response becomes maximal in approximately 20–30 min. The display is red–far-red reversible; this means that after the new higher growth rate has been established by additional far-red the growth rate can be reduced by additional red. Apart from satisfying the operational criterion of red–far-red reversibility, this result indicates that the growth responses are closely related to whatever is the primary action of phytochrome. The responses are best seen as a continuous and reversible modulation of growth rate by a red:far-red ratio operating somehow through phytochrome.

The transducer methodology allows growth rate to be related to $[P_{fr}]$ because, during the short period between light treatment and measurement of the response, it is unlikely that there is a substantial change in the total amount of phytochrome present. If the increases in growth rate caused by added far-red are plotted against P_{fr}/P_{total} measured *in vitro* by using phytochrome samples exposed to the light sources in exactly the same geometry as the irradiated plant tissues, the relation shows extreme sensitivity to a small decrease in P_{fr}/P_{total}. If different wavelengths of additional far-red are used (i.e. 700, 719, 739 nm) the increases in growth rate can be seen to lie on the same curve when plotted against P_{fr}/P_{total} (Morgan *et al.* 1980, 1981). These three additional far-red wavelengths should increase phytochrome cycling to different degrees, and yet they establish the same relation between growth rate and P_{fr}/P_{total}. In another experiment to assess the possible influence of phytochrome cycling, different fluence rates of background white light were used; again the increments in growth rate caused by the far-red could be plotted on the same curve against P_{fr}/P_{total}. In all experiments we have done using the transducer the extreme sensitivity to a small decrease in P_{fr}/P_{total} was observed.

To account for the responses of light-grown plants it is necessary to explain:

(*a*) the extreme sensitivity to a small decrease in P_{fr}/P_{total} at high values of P_{fr}/P_{total} observed when the growth effects are measured continuously;

(*b*) the apparent lack of this extreme sensitivity when the growth effects are measured over a long time;

(*c*) the linear relation between elongation and P_{fr}/P_{total} observed with long-term measurements of growth.

On points (*a*) and (*b*) it is possible that changes in the amounts of P_{total} over the long times used might account for the apparent lack of extreme sensitivity to far-red seen on a long-term basis. We have been unable in our laboratory to measure the phytochrome content of light-grown mustard seedlings *in vivo*, even when treated with the bleaching herbicide Norflurazon; however, we have done similar experiments with maize seedlings and the level of total phytochrome has been shown to increase in seedlings exposed over several days to white light containing a relatively high proportion of far-red (i.e. low red:far-red ratio), whereas seedlings exposed to white light with a high red:far-red ratio showed a small decrease in P_{total} (Smith 1981). In these experiments the growth of the maize seedlings was shown to be linearly related to P_{fr}/P_{total} even after 48–72 h of exposure to light, at which point $[P_{fr}]$ was equal in all four treatments given; in this example therefore there was no obvious relation between $[P_{fr}]$ and growth,

The most difficult point to reconcile between the responses of the light-grown plant and those of the dark-grown plant is the form of the relation between response and P_{fr}/P_{total} observed when the growth effects are measured by transducer, i.e. under conditions in which P_{total} should not change. Formally, the relations observed in dark-grown plants (described earlier) could not account for the shape of this curve, with a high sensitivity to a small change in P_{fr}/P_{total} at high P_{fr}/P_{total} levels. Indeed on the argument advanced above in relation to the very early work on phytochrome, one would be justified in concluding that the production of a small amount of P_r leads to an amplified effect on growth, and that in light-grown plants it is P_r rather than P_{fr}, that is active! It does seem, however, that a useful generalization may be that dark-grown plants are exceptionally sensitive to a small change in P_{fr}/P_{total} at low P_{fr}/P_{total} levels, whereas light-grown plants are exceptionally sensitive to a small change in P_{fr}/P_{total} at high P_{fr}/P_{total} levels.

MODELS

There is no shortage of models for phytochrome action but it seems that none of the current models is capable of accounting for all the responses outlined in this article. Strangely enough, none of the currently more fashionable models considers P_{fr} to be the *only* active form. The model of Schäfer (1975) adheres most closely to the central concept but even here it was found necessary to postulate two different forms of P_{fr}, one responsible for the red–far-red inducible responses and the other for the high-irradiance responses. It is difficult to incorporate into Schäfer's model the extreme sensitivity to small amounts of far-red exhibited by the light-grown plant. The other model most actively discussed at present is that of Johnson & Tasker (1979). In this model, action of phytochrome in the red–far-red reversible induction responses is considered to be a direct function of P_{fr}, whereas action in the high-irradiance response is due to an interaction between P_{fr} and a substance X, of which the level, or availability, is determined by some cycling-driven process. Again it is difficult to build into this model a very high sensitivity to small changes in P_{fr}/P_{total} at high values of P_{fr}/P_{total}. Very recently, VanDerWoude (1982, 1983) has developed a model of phytochrome action based upon the proposal that phytochrome exists *in vivo* as a dimer in which each monomer has a single chromophore. The model was developed to account for the extreme sensitivity of lettuce seeds, previously given a high-temperature treatment before germination, to very small amounts of red light. The model proposes that the extreme sensitivity to red light is due to the conversion of one of the pair of chromophores in the dimer to P_{fr}; at levels of red light that convert both chromophores to P_{fr} the sensitivity becomes reduced. Even in this model, however, activity is not considered to be strictly a function of $[P_{fr}]$, with P_r–P_{fr} considered to have different activity from that of P_{fr}–P_{fr}. It may be possible to build into the VanDerWoude model a high sensitivity to small amounts of far-red, if this would convert P_{fr}–P_{fr} into P_{fr}–P_r, but it would be necessary to assume that a small amount of P_{fr}–P_r would exert a marked effect in the presence of relatively very large amounts of P_{fr}–P_{fr}.

One way of circumventing the awkward fact that spectrophotometrically measurable $[P_{fr}]$ does not always correlate with response is to postulate the existence of (at least) two populations of P_{fr}, one present in small amounts and responsible for the growth effects, and the other present in much larger amounts and therefore measurable spectrophotometrically, but physiologically inactive. This idea has been repeatedly adduced in attempts to resolve the paradoxes between response and $[P_{fr}]$ (see Hillman (1967, 1972) for reviews) and is currently popular once more. Evidence is accumulating that two populations of phytochrome exist that differ in their susceptibility to degradation, such that most of the phytochrome in the dark-grown plant is of the unstable variety, whereas that in light-grown plants is relatively very stable (Brockmann & Schäfer 1982; see also the papers by Furuya, Mohr & Schäfer, and Quail *et al.* in this symposium). Indeed, the view is developing that the unstable phytochrome present in large amounts in etiolated plants serves an 'antennae' function, enabling the seedling under the soil to detect the soil surface with great sensitivity (see Quail *et al.* this symposium), whereas the stable phytochrome of the mature plant operates essentially as a detector of red:far-red ratio (see Smith 1982). Although the second part of this attractive concept is quite soundly based, much research will be needed to establish the antenna function proposal.

Blaauw-Jansen (1983) has recently suggested that P_{fr} in dark-grown seedlings is stable until a critical concentration is reached (*ca.* 2 % P_{total} in *Avena*), above which P_{fr} degradation occurs as a first-order reaction. Such behaviour is formally equivalent to two populations, one stable and one unstable and, it is claimed, can account for the very high sensitivity of etiolated seedlings to small amounts of red light.

Although the concept of two populations of phytochrome may possibly reconcile the paradoxes between the physiology and the spectrophotometry, it should be borne in mind that taking refuge in such a concept also renders highly questionable those cases in which measured $[P_{fr}]$ appears to correlate positively with response (cf. Smith 1975, p.107). If there are two populations of P_{fr} in the one case, why should there not also be two populations in the other?

On the face of it none of the current models, nor any of the relations observed in the dark-grown plant, appear to account for the relations observed in the light-grown plant. Obviously much more information will be necessary before we can integrate all the various relations between response and phytochrome parameters, and this is particularly true for the light-grown plant where as yet very few intensive investigations have been mounted. It does seem, however, as if it will be necessary to construct a model for phytochrome action that allows for extreme sensitivity to low $[P_{fr}]$ in the dark-grown plant and extreme sensitivity to small changes in $[P_{fr}]$ at high $[P_{fr}]$ in the light-grown plant. It clearly would be very difficult to do this if one adheres to the limiting concept that P_{fr} is the *only* active component of phytcchrome.

Conclusions

The concept that P_{fr} is the only active form of phytochrome is certainly an attractive hypothesis, but equally certainly it is not based on a critical and unbiased view of the available evidence. This evidence is basically correlative in nature and there are sufficient examples of lack of correlation between $[P_{fr}]$ and response to indicate to the cautious mind that scepticism should be in order. Lack of correlation can be seen in the responses of dark-grown plants, in the high-irradiance responses, and in the responses of light-grown plants. Furthermore, even in cases where $[P_{fr}]$ appears to be correlated with response the relations are not uniform and indeed some of the relations are difficult to interpret. Particular emphasis should be placed upon the observations of extreme sensitivity to low levels of $[P_{fr}]$ in the dark-grown plant and dark-imbibed seeds, and the equally extreme sensitivity to small amounts of added far-red light in plants growing in white light establishing a high P_{fr}/P_{total}. It can only be concluded that the present models for phytochrome action are inadequate to account for all the various relations that appear to exist, and one is forced to the conclusion that P_{fr} is unlikely to be the *only* active form of phytochrome.

REFERENCES

Bartley, M. R. & Frankland, B. 1982 Analysis of the dual role of phytochrome in the photoinhibition of seed germination. *Nature, Lond.* **300**, 750–752.

Beggs, C. J., Holmes, M. G., Jabben, M. & Schäfer, E. 1980 Action spectra for the inhibition of hypocotyl growth by continuous irradiation in light- and dark-grown *Sinapis alba* L. seedlings. *Pl. Physiol.* **66**, 615–618.

Blaauw-Jansen, G. 1983 Thoughts on the possible role of phytochrome destruction in phytochrome-controlled responses. *Pl. Cell Envir.* **6**, 173–179.

Borthwick, H. A., Hendricks, S. B., Parker, M. W., Toole, E. H. & Toole, V. K. 1952 A reversible reaction controlling seed germination. *Proc. natn. Acad. Sci. U.S.A.* **38**, 662–666.

Borthwick, H. A., Hendricks, S. B., Toole, E. H. & Toole, V. K. 1954 Action of light on lettuce-seed germination. *Bot. Gaz.* **115**, 205–225.

Briggs, W. R. & Chon, H. P. 1966 The physiological versus the spectrophotometric status of phytochrome in corn coleoptiles. *Pl. Physiol.* **41**, 1159–1166.

Brockmann, J. & Schäfer, E. 1982 Analysis of P_{fr} destruction in *Amaranthus caudatus* L. – evidence for two pools of phytochrome. *Photochem. Photobiol.* **35**, 555–558.

Chon, H. P. & Briggs, W. R. 1966 Effects of red light on the phototropic sensitivity of corn coleoptiles. *Pl. Physiol.* **41**, 1715–1724.

Drumm, H. & Mohr, H. 1974 The dose response curve in phytochrome-mediated anthocyanin synthesis in the mustard seedling. *Photochem. Photobiol.* **20**, 151–157.

Gammerman, A. Ya. & Fukshansky, L. 1974 A mathematical model of phytochrome – the receptor of photomorphogenetic processes in plants. *Ontogenez.* **5**, 122–129.

Hartmann, K. M. 1966 A general hypothesis to interpret 'high energy phenomena' of photomorphogenesis on the basis of phytochrome. *Photochem. Photobiol.* **5**, 349–366.

Hartmann, K. M. 1967a Phytochrome 730 (P_{fr}), the effector of the 'high energy photomorphogenic reaction' in the far-red region. *Naturwissenschaften* **54**, 544.

Hartmann, K. M. 1967b Ein Wirkungsspektrum der Photomorphogenese unter Hochenergiebedingungen und seine interpretation auf der Basis des Phytochroms (Hypokotylwachstumshemmung bei *Lactuca sativa* L.). *Z. Naturf.* **22b**, 1172–1175.

Heim, B. & Schäfer, E. 1982 Light-controlled inhibition of hypocotyl growth in *Sinapis alba* L. seedlings. Fluence-rate dependence of hourly light pulses and continuous irradiation. *Planta* **154**, 150–155.

Heim, B. & Schäfer, E. 1983 Hypocotyl growth in dark-grown *Sinapis alba* L.: fluence-rate dependence of continuous and hourly pulsed irradiation with far-red. *Pl. Cell Envir.* **6**. (In the press.)

Hillman, W. S. 1965 Phytochrome conversion by brief illumination and the subsequent elongation of isolated *Pisum* stem segments. *Physiologia Pl.* **18**, 346–358.

Hillman, W. S. 1966 Responses of *Avena* and *Pisum* tissues to phytochrome conversion by red light. *Pl. Physiol.* **41**, 907–908.

Hillman, W. S. 1967 The physiology of phytochrome. *A. Rev. Pl. Physiol.* **18**, 301–324.

Hillman, W. S. 1972 On the physiological significance of *in vivo* phytochrome assays. In *Phytochrome* (ed. K. Mitrakos & W. Shropshire Jr), pp. 573–584. New York and London: Academic Press.

Holmes, M. G. & Schäfer, E. 1981 Action spectra for changes in the 'high irradiance reaction' in hypocotyls of *Sinapis alba* L. *Planta* **153**, 267–272.

Holmes, M. G. & Smith, H. 1975 The function of phytochrome in plants growing in the natural environment. *Nature, Lond.* **254**, 512–514.

Holmes, M. G. & Smith, H. 1977 The function of phytochrome in the natural environment. IV. Light quality and plant development. *Photochem. Photobiol.* **25**, 551–557.

Johnson, C. B. 1980 The effect of red light in the high irradiance reaction of phytochrome: evidence for an interaction between P_{fr} and a phytochrome cycling-driven process. *Pl. Cell. Envir.* **3**, 45–51.

Johnson, C. B. & Tasker, R. 1979 A scheme to account quantitatively for the action of phytochrome in etiolated and light-grown plants. *Pl. Cell Envir.* **2**, 259–265.

Loercher, L. 1966 Phytochrome changes correlated to mesocotyl inhibition in etiolated *Avena* seedlings. *Pl. Physiol.* **41**, 932–936.

Mandoli, D. F. & Briggs, W. R. 1981 Phytochrome control of two low-irradiance responses in etiolated oat seedlings. *Pl. Physiol.* **67**, 733–739.

Mohr, H. 1972 *Lectures on photomorphogenesis.* Berlin, Heidelberg and New York: Springer-Verlag.

Mohr, H., Drumm, H., Schmidt, R. & Steinitz, B. 1979 The effect of light pretreatments on phytochrome-mediated induction of anthocyanin and phenylalanine ammonia-lyase. *Planta* **146**, 369–376.

Morgan, D. C., Child, R. & Smith, H. 1981 Absence of fluence rate dependency of phytochrome modulation of stem extension in light-grown *Sinapis alba* L. *Planta* **353**, 497–498.

Morgan, D. C., O'Brien, T. & Smith, H. 1980 Rapid photomodulation of stem extension in light-grown *Sinapis alba* L. *Planta* **150**, 95–101.

Morgan, D. C. & Smith, H. 1976 Linear relationship between phytochrome photoequilibrium and growth in plants under simulated natural radiation. *Nature, Lond.* **262**, 210–212.

Morgan, D. C. & Smith, H. 1978 The relationship between phytochrome photoequilibrium and development in light-grown *Chenopodium album* L. *Planta* **142**, 187–193.

Morgan, D. C. & Smith, H. 1979 A systematic relationship between phytochrome-controlled development and species habitat, for plants grown in simulated natural radiation. *Planta* **145**, 253–258.

Oelmüller, R. & Mohr, H. 1983 Responsivity amplification by light in phytochrome-mediated induction of chloroplast glyceraldehyde-3-phosphate dehydrogenase (NADP-dependent, EC 1.2.1.13) in the shoot of milo (*Sorghum vulgare* Pers.). *Pl. Cell Environ.* (In the press.)

Oelze-Karow, H. & Mohr, H. 1970 Phytochrome-mediated repression of enzyme synthesis (lipoxygenase): a threshold phenomenon. *Proc. natn. Acad. Sci. U.S.A.* **65**, 51–57.

Oelze-Karow, H. & Mohr, H. 1973 Quantitative correlation between spectrophotometric phytochrome assay and physiological response. *Photochem. Photobiol.* **18**, 319–330.

Parker, M. W., Hendricks, S. B., Borthwick, H. A. & Scully, N. J. 1946 Action spectrum for the photoperiodic control of floral initiation of short-day plants. *Bot. Gaz.* **108**, 1–26.

Parker, M. W., Hendricks, S. B., Borthwick, H. A. & Went, F. W. 1949 Spectral sensitivities for leaf and stem growth of etiolated pea seedlings and their similarity to action spectra for photoperiodism. *Am. J. Bot.* **36**, 194–204.

Raven, C. W. & Shropshire, W. Jr 1975 Photoregulation of logarithmic fluence-response curves for phytochrome control of chlorophyll formation in *Pisum sativum* L. *Photochem. Photobiol.* **21**, 423–429.

Schäfer, E. 1975 A new approach to explain the 'high irradiance response' of photomorphogenesis on the basis of phytochrome. *J. math. Biol.* **2**, 41–56.

Schäfer, E. & Mohr, H. 1974 Irradiance dependency of the phytochrome system in cotyledons of mustard (*Sinapis alba* L.). *J. math. Biol.* **1**, 9–15.

Schmidt, R. & Mohr, H. 1981 Time-dependent changes in the responsiveness to light of phytochrome-mediated anthocyanin synthesis. *Pl. Cell Envir.* **4**, 433–437.

Schmidt, R. & Mohr, H. 1982 Evidence that a mustard seedling responds to the amount of P_{fr} and not to the $P_{fr}:P_{total}$ ratio. *Pl. Cell Envir.* **5**, 495–499.

Schmidt, R. & Mohr, H. 1983 Time course of signal transduction in phytochrome-mediated anthocyanin synthesis in mustard cotyledons. *Pl. Cell Envir.* **6**, 235–238.

Small, J. G. C., Spruit, C. J. P., Blaauw-Jansen, G. & Blaauw, O. H. 1979 Action spectra for light-induced germination in dormant lettuce seeds. *Planta* **144**, 125–131.

Smith, H. 1975 *Phytochrome and photomorphogenesis*. London: McGraw-Hill.

Smith, H. 1981 Evidence that P_{fr} is not the active form of phytochrome in light-grown maize. *Nature, Lond.* **293**, 163–165.

Smith, H. 1982 Light quality, photoperception, and plant strategy. *A. Rev. Pl. Physiol.* **33**, 481–518.

Steinitz, B., Schäfer, E., Drumm, H. & Mohr, H. 1979 Correlation between far-red absorbing phytochrome and response in phytochrome-mediated anthocyanin synthesis. *Pl. Cell Envir.* **2**, 159–163.

VanDerWoude, W. J. 1982 Mechanism of photothermal interactions in phytochrome control of seed germination. In *Strategies of plant reproduction* (BARC Symposium no. 6) (ed. W. J. Meudt), pp. 135–143). New Jersey: Allanheld, Osmun.

VanDerWoude, W. J. 1983 A dichromophoric model for the action of phytochrome: evidence from photothermal interactions in lettuce seed germination. *Proc. natn. Acad. Sci. U.S.A.* (In the press.)

VanDerWoude, W. J. & Toole, V. K. 1980 Studies of the mechanism of enhancement of phytochrome-dependent lettuce seed germination by prechilling. *Pl. Physiol.* **66**, 220–224.

Vince-Prue, D. 1977 Photocontrol of stem elongation in light-grown plants of *Fuchsia hybrida*. *Planta* **133**, 149–156.

Whitelam, G. C. & Johnson, C. B. 1981 Temporal separation of two components of phytochrome action. *Pl. Cell Envir.* **4**, 53–57.

Phil. Trans. R. Soc. Lond. B **303**, 453–465 (1983)
Printed in Great Britain

Photocontrol of extension growth: a biophysical approach

By D. J. Cosgrove

*Department of Biology, 202 Buckhout Laboratory, Pennsylvania State University,
University Park, Pennsylvania 16802, U.S.A.*

The analysis of plant growth as a physical process is briefly reviewed. Growth requires the coordinated uptake of water and the irreversible expansion of the cell wall. Any agent that affects the growth rate must act on one or more of the parameters governing water absorption (e.g. the hydraulic conductivity or the difference in osmotic pressure of the cell contents and the water source) or cell wall expansion (e.g. wall extensibility or the yield threshold). When the hydraulic conductivity of the pathway for water transport is small enough to impede the rate of cell enlargement, a substantial gradient in water potential within the growing tissue will develop to sustain the absorption of water. In such a case, the analysis shows that turgor pressure is a key indicator for determining whether an agent acts predominantly on the osmotic properties of the tissue or on the cell wall properties. Furthermore, the dynamic response to a slight perturbation from steady-state conditions is shown to be a function of parameters for both the water relations and cell wall expansion of the tissue.

Blue irradiation of etiolated seedlings causes a large inhibition of stem elongation with lag times as short as 30 s and half-times as short as 20 to 25 s. The biophysical mechanism of blue-light suppression of growth was studied in cucumber and sunflower seedlings by means of direct and indirect measurements of turgor pressure. The results indicate that (*a*) blue light suppresses growth by influencing the cell wall properties of the growing tissue, and (*b*) the hydraulic conductivity of the growing tissue is large enough for it not to limit the rate of cell enlargement.

Introduction

The transition of a plant from dark to light conditions is followed by a number of pronounced changes in the development of the plant. Leaves expand very little in the dark, but upon onset of irradiation they unfold and expand to their full size. Just the opposite action occurs in stems where the high rate of elongation in etiolated seedlings is greatly inhibited by light. These morphogenetic developments are accompanied by many other changes at the cell, organelle, and biochemical levels. In this report I wish to focus on efforts to study the mechanism by which light alters the growth rate of shoots, with particular emphasis on the physical approach to this problem.

The effect of light on shoot growth has been the subject of numerous studies. Since the discovery of auxin and other growth regulators, many investigations have dealt with the hypothesis that light alters growth by the mediation of one or more of the plant growth regulator systems (Black & Vlitos 1972). It is indeed clear by now that light has effects on the levels of endogenous auxin and gibberellins (Briggs 1963; Iino 1982; Murray & Acton 1974), on the levels of growth antagonists (Franssen & Bruinsma 1982), on the transport of hormones through plant tissues (Pickard & Thimann 1964; Sherwin & Furuya 1973; Thornton & Thimann 1967), on the number of auxin receptor sites (Walton & Ray 1981) and on the sensitivity of tissues to exogenously applied hormones (Kende & Lang 1964; Russel & Galston

1969). What is not clear is to what degree these various biochemical responses are incidental to the change in growth rate or fully account for the light growth response. Indeed, there is disagreement about whether endogenous hormone levels normally exert a controlling influence on the growth rate of a plant tissue (Trewavas 1981). It is a fundamental problem with this experimental approach that it is difficult to quantitatively relate a particular biochemical change induced by irradiation to the growth response. As an example, some workers dispute whether the amount of auxin redistribution during phototropism of corn coleoptiles is sufficient to account for the observed change in growth rate (Firn & Digby 1980; Hall et al. 1980). One important reason for this controversy is that auxin itself does not directly control growth; rather, it acts on one or more cellular parameters that govern the growth rate.

A way around this dilemma is to investigate growth from a biophysical perspective, in which the physical parameters directly responsible for growth are measured. This approach offers the possibility of quantitatively accounting for the entire growth response in terms of changes in four biophysical parameters that govern the growth rate of a tissue.

PLANT GROWTH AS A PHYSICAL PROCESS

The growth of plant organs entails the irreversible enlargement of cells. Cleland & Rayle (1977) have calculated that if the stem cells of a redwood tree simply duplicated themselves without undergoing the usual phase of prolonged cell enlargement, the tallest redwood would stand less than two feet in height! For irreversible cell enlargement to occur, two fundamental processes are required. The cells must take up water to generate the additional cell volume, in that growing cells are more than 90 % water; and the cell wall surrounding the protoplast must irreversibly expand to accommodate the absorbed water. Conceptually, water uptake and irreversible wall expansion are distinct processes, but in growing tissues they are tightly linked with one another. Each process will be considered separately below, and then their interdependence will be discussed.

Water absorption during plant growth

As far as has been established, water transport is a passive physical process that (with certain important restrictions) occurs down a gradient in chemical potential, or equivalently, water potential. The rate of water uptake by a cell, given any specific water potential difference between its contents and the environment, depends on the hydraulic conductivity of the pathway for water transport (L), such that

$$dV_w/dt = L(\Delta\psi) = L(P-\Delta\pi),\tag{1}$$

where dV_w/dt is the rate of water uptake, $\Delta\psi$ is the difference in water potential between the cell and the water source, P is the turgor pressure of the cell, and $\Delta\pi$ is the difference in osmotic pressure between the cell contents and the water source. To a first approximation, dV_w/dt is equal to the growth rate of the cell (assuming zero transpiration). Equation (1) assumes an ideal semipermeable membrane.

When growth of multicellular tissues is considered, the mathematical treatment of water transport becomes rather more complicated (Molz & Boyer 1978; Molz & Ferrier 1982). The hydraulic conductivity of both the cell wall pathway and the cell-to-cell pathway for water transport must be considered (figure 1). Providing that conditions of 'local equilibrium' obtain (i.e. water in the cell wall space is in equilibrium with water in the nearest cells), water transport will have diffusion-type kinetics. In such a case, there will develop within a growing tissue a gradient in water potential that is induced by the expansion of the cell wall and that at the same time sustains the absorption of water by the growing cells. Models of elongation in a cylindrical piece of tissue such as a stem predict that the water potential gradient would be

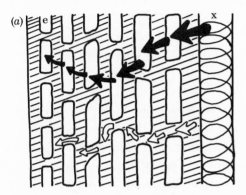 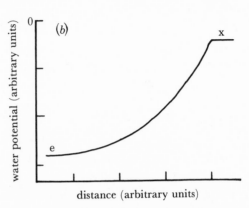

FIGURE 1. (a) Diagram showing the pathway for water flow during the growth of a multicellular tissue. Water may move from the xylem (x) toward the epidermis (e) via the cell-to-cell pathway (solid arrows) and the cell wall pathway (open arrows). (b) Theoretical water potential profile in a growing cylinder of tissue such as a stem. The profile is drawn radially from the xylem (right side) to the epidermis (left side). The size of the gradient depends on the growth rate of the tissue and the diffusivity of the tissue for water flow. (Redrawn after Molz & Boyer (1978).)

steep near the water source (the xylem) and flatter in the regions further from the xylem (see figure 1). The cells closest to the xylem must transport the water that is absorbed during their own growth as well as that absorbed by cells lying further from the centre. For this reason the gradient in water potential is steeper in this region to support the higher rate of water flow. Toward the outside of the tissue, the gradient is flatter because less water is transported.

Cell wall expansion during plant growth

In non-growing plant tissue, cells out of osmotic equilibrium would absorb water and consequently increase in turgor pressure until $\Delta\psi$ became zero (i.e. $P = \Delta\pi$), at which point net water flow would stop. However, in a growing cell, the cell walls are loosened in some fashion and the cell wall yields to the stress generated by the pressure of the cell contents. This irreversible stretching or deformation of the cell wall dissipates P (because water has a low compressibility) and hence sustains the gradient in water potential necessary for water absorption. Thus growth begins with stress relaxation of the cell wall (i.e. wall loosening), which concommittantly brings about relaxation of the turgor pressure and a lowering of the water potential of the cells. The depressed water potential permits uptake of water, which generates additional cell volume and expands the cell wall. If stress relaxation of the cell wall is stopped, then water absorption will cause the stress in the cell wall and its opposing force (turgor pressure) to increase until osmotic equilibrium is re-established.

Wall expansion is thought to be a rheological process in which the cell wall is physically distended by the forces in the wall generated by turgor pressure. A number of studies (Lockhart 1965; Cleland 1967; Green et al. 1971) have suggested that the rate of cell-wall expansion is proportional to the amount by which turgor pressure exceeds a certain minimum value, such that

$$dV_{cw}/dt = m(P - Y), \tag{2}$$

where Y is the minimum turgor pressure necessary for growth (the yield threshold), m is a parameter termed the cell wall extensibility, and dV_{cw}/dt is the rate of expansion of cell wall, expressed in volumetric units. It should be emphasized here that the term 'cell wall extensibility' as used in this context does not refer to the elastic extensibility measured by various mechanical means, but refers to the coefficient relating the rate of irreversible cell wall expansion to the 'effective' turgor pressure $(P - Y)$. Equation (2) is an empirically derived relation and is somewhat deceptive in that m and Y are not simple physical parameters describing time-independent rheological properties of the cell wall (Green *et al.* 1971; Green & Cummins 1974; Ray & Ruesink 1962). Rather, they seem to depend on the rate of cellular or biochemical processes that loosen the cell wall and are coupled to the respiration rate (i.e. they are energy dependent). The nature of the wall-loosening and subsequent expansion processes have yet to be established and the physical meanings of m and Y are not clear, except as operationally defined in (2). It is probable that they represent the result of a complex cellular process and it thus may be misleading to represent this process in such a simple fashion. Further insight into the physical and biochemical nature of wall extension and its relation to turgor pressure is vitally needed.

In contrast, the water transport equation has a strong basis in the theory of irreversible thermodynamics and is well supported experimentally (Molz & Ferrier 1982; Dainty 1976). The parameters of (1) are well defined and may be measured in a number of independent ways. Of course, the values of L and $\Delta\pi$ depend upon the maintenance of membrane integrity and solute accumulation, which are uphill processes energetically and thus in the long term also depend on metabolic energy, although only indirectly.

Predictions of the physical theory

Under conditions of steady-state growth, the rates of water uptake and irreversible wall expansion are assumed to be equal to each other and we may combine (1) and (2) to obtain the general equation governing the rate of plant growth (dV/dt):

$$dV/dt = Lm(\Delta\pi - Y)/(L + m). \tag{3}$$

The growth rate (dV/dt) thus depends on the parameters controlling both water transport and irreversible wall expansion. These two processes are linked by the common dependent variable, turgor pressure, which is eliminated to obtain (3). Although turgor pressure does not explicitly appear in (3), an understanding of its role in coordinating wall expansion and water absorption is crucial for an understanding of cell expansive growth.

If water transport is a very rapid process in comparison with cell wall expansion (more specifically, if the hydraulic conductivity is much greater than the wall extensibility), the turgor pressure (and water potential) in a growing cell is negligibly affected by the expansion of the cell wall (Cosgrove 1981 a). Water can enter the cell quickly enough to maintain the turgor pressure (almost) at the equilibrium value $(\Delta\pi)$. Consequently the hydraulic conductivity cannot exert a controlling influence on the rate of growth; the yielding properties of the cell wall are then of paramount importance. On the other hand, if water uptake is a slow process, wall expansion may occur at a sufficiently rapid rate to dissipate turgor pressure significantly (figure 2). The lower turgor pressure in turn decreases the rate of wall expansion (equation (2)) and the hydraulic conductivity in this case may exert a significant control on the growth rate by influencing steady-state turgor pressure.

As indicated by this analysis, the process of wall expansion tends to decrease turgor pressure, whereas the process of water uptake tends to increase it. In principle, the turgor pressure of

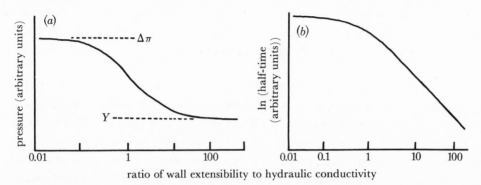

ratio of wall extensibility to hydraulic conductivity

FIGURE 2. (a) Turgor pressure as a function of the ratio m/L. When wall extensibility (m) is much smaller than hydraulic conductivity (L), the tissue is essentially at osmotic equilibrium and $P = \Delta\pi$. At the other extreme, turgor pressure is dissipated by rapid expansion of the cell wall, so that $P = Y$. Within the range 0.1–10, the turgor pressure is determined by the balance point between water uptake and wall expansion. (b) Half-time of tissue as a function of the ratio m/L. Only when the value of wall extensibility approaches or exceeds the value of hydraulic conductivity is the half-time affected by the cell wall properties. Note the logarithmic scales. (After Cosgrove (1981 a).)

a growing cell must be less than that of an equivalent non-growing cell; it must lie somewhere in the range between $\Delta\pi$ and Y (see figure 2), the exact value being determined by the balance point between wall expansion and water uptake (Cosgrove 1981 a). In single isolated cells, water transport is so rapid that the cell is essentially in osmotic equilibrium with its bathing solution (Cosgrove 1981 a). In multicellular tissue, however, it is possible for the hydraulic conductivity to decrease sufficiently (because of the long pathway for transport) for the cell layers remote from the water source to be far from osmotic equilibrium (Molz & Boyer 1978). Indeed, several reports support this view in that they show that growing tissues of various plant organs have low water potentials (Ray & Ruesink 1963; Boyer 1968; Molz & Boyer 1978; Michelena & Boyer 1982).

In such tissues we can predict that if an agent inhibits growth, it ought to change the steady-state balance point of turgor pressure upwards or downwards (toward $\Delta\pi$ or Y), depending on whether it acts primarily on the water transport characteristics of the tissue (L, $\Delta\pi$) or on the yielding properties of the cell wall (m, Y). Turgor pressure in principle should increase if the agent acts on the cell wall properties, but decrease if the agent acts on the osmotic properties (Cosgrove 1981 a). The *amount* by which turgor pressure changes depends on the size of the span between $\Delta\pi$ and Y, and whether the balance point is already very close to $\Delta\pi$ or Y (see figure 2). For example, if the growth rate were controlled entirely by the wall properties (i.e. if hydraulic conductivity were much larger than wall extensibility), the change in turgor pressure brought about by a change in wall extensibility would be very small. Turgor pressure would be most responsive when wall extensibility and hydraulic conductivity were equal in magnitude.

A second prediction that may be derived from the physical analysis of plant growth concerns the time course for re-establishment of the steady state after a slight perturbation (Cosgrove 1981 a). Consider first a non-growing cell. If the cell is brought out of osmotic equilibrium with an external solution (either by changing the water potential of the solution or the water potential of the cell contents), water will be induced to flow across the cell membranes until equilibrium is re-established. The rate of water flow will be large at first and will decrease exponentially as the equilibrium point is reached. The half-time for the transient flow of water is determined entirely by the physical parameters governing water transport (the hydraulic conductivity, the osmotic pressure of the cell contents, the volumetric elastic modulus of the cell, and the cell geometry) (see Dainty 1976; Zimmermann & Steudle 1978; Molz & Ferrier 1982).

Now consider a growing cell. The water potential of the cell is affected not only by the process of water transport, but also by the process of cell wall expansion. If wall expansion is slow (such that the growing cell is close to osmotic equilibrium), the half-time for the re-establishment of steady state after a perturbation is hardly affected by the wall expansion process; the half-time will be the same as in a non-growing cell. If wall expansion is more rapid, i.e. if wall extensibility is similar to or larger than hydraulic conductivity, the half-time will be much faster for the growing cell (see figure 2). This half-time will be governed not only by the water relations parameters, but also the parameters controlling wall expansion, as shown by the analysis of Cosgrove (1981 a).

These two predictions based on the physical theory of plant growth have been used to investigate the inhibition of elongation growth by blue irradiation.

CASE STUDY: BLUE LIGHT INHIBITION OF GROWTH

Irradiation of dark-grown dicotyledonous seedlings with blue light induces a very rapid decrease in the elongation rate of the stem (Meijer 1968; Gaba & Black 1979). The latent period between the start of irradiation and the start of the growth inhibition is as short as 20 s in some species; a more typical value is about 60 s (Cosgrove 1981 b). A large response (say 25–75 % inhibition) requires that the growing tissue itself be irradiated with a high fluence rate of blue light (1–5 W m^{-2} are typical values; see Cosgrove (1981 b) and Cosgrove & Green (1981)). It has been shown from a number of criteria that this blue-light response is mediated by a specific blue light photoreceptor, distinct from phytochrome (Cosgrove 1982; Gaba & Black 1979; Holmes & Schäfer 1981; Thomas & Dickinson 1979).

The interesting aspect of this light response, from a growth physiologist's point of view, is that it occurs so rapidly. In certain species the growth inhibition has approximately the form of an exponential decay in the growth rate (see figure 3; see also Cosgrove (1981 b) and Cosgrove & Green 1981). In cucumber the half-time for the inhibition is between 15 and 30 s. Thus the full growth inhibition is complete in less than 5 min after the start of the irradiation. This is an important advantage for studying the mechanism of this growth response, in that little time is available for secondary responses to complicate the investigation. In contrast, red-light induced inhibitions of growth generally have longer lag times (Meijer 1968; Morgan et al. 1980; Vanderhoef & Briggs 1978) and probably involve multiple response mechanisms (Iino 1982).

The theoretical model of plant growth discussed above ignored the fact that transpirational water loss may also influence growth. For example, if blue light stimulated the opening of stomata in the epidermis of the stem or leaves, this might increase transpiration enough to cause the turgor pressure, and consequently the growth rate, to fall. This mechanism must be considered because blue light has been shown to stimulate stomatal opening in leaves (Zieger *et al.* 1981). To test this hypothesis, cucumber seedlings were mounted in a chamber that permitted the growing region of the stem to be submerged in aerated water. The growth rate

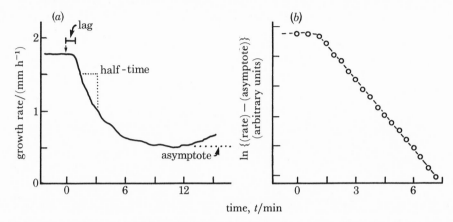

FIGURE 3. Time course of growth inhibition by blue light. (*a*) At the point indicated by the arrow, a dark-grown sunflower seedling was irradiated with 16 s of blue light (2.6 W m^{-2}). After a lag of *ca.* 1 min, the growth rate declined in an approximately exponential fashion to a low value (asymptote). (*b*) The half-time of the inhibition may be calculated from the slope of the data replotted as ln (rate minus asymptotic rate) against time. (Redrawn from Cosgrove & Green (1981).)

was measured continuously with a displacement transducer as described elsewhere (Cosgrove 1982). When a stable growth rate in the dark was attained, the seedling was irradiated with blue light (3 W m^{-2}). If the hypothesis of stomatal opening were true, blue light should have had no effect on the growth rate of such submerged seedlings, where transpiration was essentially eliminated. In fact, a typical growth inhibition was observed (data not shown). Thus blue light must act by altering one or more of the parameters in (3), e.g. $\Delta\pi$, L, m or Y.

A possible change in the concentration of intracellular solutes by blue irradiation was examined by measuring the osmolality of expressed cell sap from the growing region of cucumber seedlings exposed to blue light (2 W m^{-2}) for various lengths of time. As shown in figure 4, no effect of blue light was observed; blue light does not cause changes in bulk π.

In view of the rapid exponential kinetics of the light-growth response, it is possible that blue light causes the plasmalemma to become 'leaky' to intracellular solutes. In such a case, solutes would build up in the cell-wall free space, decreasing the gradient in osmotic pressure across the cell membrane and consequently decreasing turgor pressure and growth rate. Because free space water constitutes less than 5 % of the total volume of the tissue (Cosgrove & Cleland 1983*a*), such leakage might involve the leakage of only a very small proportion of the intracellular solutes. Furthermore, cell sap expressed from stem tissue represents a volume-weighted average of the combined intracellular and extracellular solutions, so that a redistribution of solutes within the tissue would not be detected by bulk π measurements in any case. Such a redistribution, however, would be detected by turgor pressure measurements.

Turgor pressure measurements

As pointed out in the theory section, turgor pressure is a key indicator for the mechanism by which an agent affects the plant growth. The analysis predicts a decrease in turgor pressure if blue light acts on the osmotic properties of the tissue, an increase if blue light acts on the cell wall. This prediction has been tested now by two different methods with similar results.

Qualitative changes in turgor pressure during the inhibition of stem elongation by blue light were first measured by using an indirect technique that measures the mechanical rigidity of

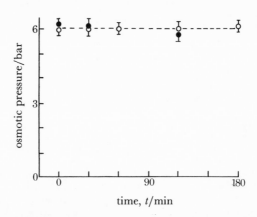

FIGURE 4. Osmotic pressure of cell sap expressed from 1 cm sections cut from the growing stem region of cucumber seedlings irradiated with blue light (2 W m^{-2}) for various length of time. Open circles are data from irradiated plants, filled circles from dark controls. Points plotted are means $\pm$ s.d. of 10 samples. (1 bar $= 10^5$ Pa.)

plant tissue (Virgin 1955). An intact stem was firmly mounted in the middle of the growing region and forced to vibrate at its resonance frequency. The more rigid the stem (i.e. the greater the elastic modulus), the higher is the resonance frequency. The cell walls of young growing tissue are very soft and weak; most of the stiffness of the tissue results from the hydrostatic pressure (turgor pressure) of the cell contents, which puts the cell wall in tension (Falk *et al.* 1958; Nilsson *et al.* 1958). In effect, such plant tissues have a hydraulic skeleton or support structure (Wainwright 1970). The rigidity of such tissue is thus largely a function of the turgor pressure of the cells. Changes in turgor pressure are reflected in changes in the resonance frequency of the stem (Virgin 1955; Nilsson *et al.* 1958).

Cucumber and sunflower seedlings were connected under dim green light to an apparatus which permitted simultaneous measurements of the growth rate and the resonance frequency of the elongating stem (Cosgrove & Green 1981). When growth was suppressed by blue light, the rigidity of the stem, measured by the resonance method, increased very slightly (figure 5). This resonance method is useful for indicating the *direction* of change in turgor pressure, but has several weaknesses. It is not very quantitative, it is sensitive to the geometry of the tissue, and it may be sensitive to changes in the mechanical properties of the cell wall (Nilsson *et al.* 1958; Uhrström 1969). Despite these limitations, we estimated that when the growth rate was inhibited 50 % by blue irradiation, the turgor pressure increased by less than 0.2 bar (2×10^4 Pa). That is, it hardly changed at all.

Recently these results were confirmed by using the pressure probe technique (Hüsken *et al.* 1978). In these experiments, the turgor pressure of an individual cell in the outer cortical region of the stem of an intact cucumber seedling was monitored at the same time that the growth rate of the whole stem was measured. When stem elongation was inhibited by blue irradiation, no change in turgor pressure was detected (Cosgrove, in preparation). The practical resolution of the pressure probe method is approximately 0.1 bar (10^4 Pa).

These results admit two interpretations. First, it could be that blue light causes a simultaneous and proportional change in both the osmotic and the cell wall properties of the tissue. Thus

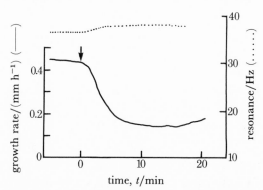

FIGURE 5. Changes in growth rate and resonance frequency in a sunflower seedling in response to a single 16 s pulse of blue light (4 W m^{-2}). The growth rate (solid line) is strongly inhibited whereas the resonance frequency of the stem (broken line) increases only very slightly. The arrow indicates when the light was given.

TABLE 1. EFFECT OF LIGHT DOSE ON GROWTH INHIBITION IN SUNFLOWER SEEDLINGS

(The growth response of seedlings to a 16 s pulse of blue light was broken down into three parameters that completely describe the response. Significance was tested by two-way analysis of variance (from Cosgrove & Green 1981).)

light dose J m^{-2}	mean asymptote† (percentage of dark rate)	mean half-time‡ s	mean lag time‡ s
144	29	141	62
40	36.5	160	55
8	45	136	65

† Significant effect of light dose at the 0.001 probability level.
‡ No significant effect.

the balance point between wall expansion and water uptake would be unaltered and the turgor pressure would remain unchanged. It would be a remarkable coincidence indeed if such different processes were affected in such a fashion. However, this hypothesis may be rejected for the following reason. The half-time of inhibition appears to be identical with the half-time for readjustment in growth rate after perturbation (Cosgrove & Green 1981). If blue light decreased both the wall extensibility and the hydraulic conductivity of the growing tissue, the half-time of the growth response would be strongly affected. However, the half-time was experimentally found to be independent of the degree of inhibition and the light dose (table 1).

The second interpretation of the turgor pressure data is that the hydraulic conductivity of the tissue is so large that it does not limit growth. In such a case the gradient in water potential that sustains growth is insignificant and the turgor pressure of the growing tissue is only negligibly lower than in an equivalent non-growing tissue. Likewise, the half-time of the tissue is unaffected by wall expansion. At present this interpretation seems most in agreement with the results of the experiments.

[115]

These data indirectly support the hypothesis that blue light acts on the cell wall properties to inhibit growth. This suggests that the yielding properties of the wall (m or Y) are under remarkably tight control by cellular processes. Two observations make it unlikely that the blue light effect is mediated by auxin. First, the lag time of the light response is as short as 20 s whereas the lag time for auxin stimulation of growth is 10–15 min. The latent period with auxin treatment evidently does not represent the time required for auxin uptake but rather the time required for the intermediate steps between auxin uptake and the change in the cell-wall yielding properties (Ray 1974). Although direct data are lacking, it is reasonable to expect that if we could somehow cause an instantaneous decrease in auxin concentration, transport or sensitivity, a minimum lag of 10–15 min would follow before a decrease in the growth rate would be seen. Thus the large disparity between the lag time for response to blue irradiation and auxin treatment (20 s compared with 10–15 min) argues against involvement of auxin in the rapid light response. Second, it has been shown by other investigators (Uhrström 1969; Burström et al. 1967; Göring et al. 1975) and confirmed in this laboratory that auxin induces a rather large decrease in the rigidity of the growing tissue (measured by the resonance frequency method). In principle, this change in rigidity might be due to a loss of turgor pressure or an *elastic* 'loosening' of the cell wall. Direct measurements with the pressure probe technique show that auxin does not affect turgor pressure over this short period (Cosgrove & Cleland 1983 b); therefore the decrease in tissue rigidity upon auxin treatment must be an effect on the mechanical characteristics of the cell wall. In contrast, blue light has only a negligible effect on the rigidity of the growing tissue. These observations taken collectively suggest that auxin and blue light affect the cell wall by different mechanisms.

The observation that rapid and large suppression of elongation by blue light has only a negligible effect on the turgor pressure of the growing tissue deserves some comment. Several studies (Ray & Ruesink 1963; Boyer 1968; Molz & Boyer 1978; Michelena & Boyer 1982) have shown that growing tissues have water potentials in the range of -2 to -5 bar (-2 to -5×10^5 Pa). These low water potentials have been thought to result from the process of cell-wall expansion driving the tissue far out of osmotic equilibrium. When the water potential of growing cucumber stem tissue was estimated from measurements of turgor pressure and osmotic pressure, water potential values of -3 to -4 bar (-3 to -4×10^5 Pa) were obtained (Cosgrove, in prep.). This observation seems to be at variance with the conclusion reached above from the blue-light studies that the water potential gradient that supports growth in cucumber stems is small. However, Cosgrove & Cleland (1983 a) have recently found that the cell-wall free space of growing stem tissues contains solute concentrations high enough to account for most of this low water potential. This finding resolves the apparent paradox posed by the conclusion that hydraulic conductivity does not limit growth and observations of low water potentials in growing tissue. This point is discussed further by Cosgrove & Cleland (1983 a, b).

The process of cell-wall loosening is poorly understood. It must involve the breakage and perhaps reformation of load-bearing bonds within the cell wall. The nature of these critical bonds has not been established, nor is it clear whether the wall may be loosened in more than one way. The acid growth hypothesis for auxin action (see Cleland & Rayle 1977) proposes that the pH of the cell wall is a major control point for cell wall expansion. But this hypothesis is far from a complete or even sufficient model of cell wall expansion and cannot explain growth induced by cytokinin and gibberellin in at least some tissues (Ross & Rayle 1982; Stuart & Jones 1978). This is an area that requires further experimental investigation.

Conclusion

The analysis of plant growth as a physical process offers a powerful framework within which the mechanism of action of any agent that alters the growth rate may be investigated. Growth is shown to be controlled by two parameters governing water uptake (hydraulic conductivity and the osmotic pressure difference between the cell contents and the water source) and two operationally defined parameters describing the yielding characteristics of the cell wall (wall extensibility and yield threshold). The analysis shows that turgor pressure is a key indicator for the mechanism by which an agent affects growth. Kinetic analysis of the re-establishment of the steady state after a perturbation in the growth rate of a tissue is another useful experimental technique.

The rapid inhibition of stem elongation by blue light was investigated within this framework. The experimental results support the conclusions that (a) blue irradiation suppresses growth by acting on the yielding characteristics of the cell wall, and (b) the hydraulic conductivity of growing cucumber stems is sufficiently large for it not to impede the rate of cell enlargement. Further research into the mechanism of stress relaxation (loosening) and expansion of the cell wall is needed.

References

Black, M. & Vlitos, A. J. 1972 Possible interrelationships of phytochrome and plant hormones. In *Phytochrome* (ed. K. Mitrakos & W. Shropshire), pp. 517–549. London: Academic Press.

Boyer, J. S. 1968 Relationship of water potential to growth of leaves. *Pl. Physiol.* **43**, 1056–1062.

Briggs, W. 1963 Red light, auxin relationships, and the phototropic responses of corn and oat coleoptiles. *Am. J. Bot.* **50**, 196–207.

Burström, H., Uhrström, I. & Wurscher, R. 1967 Growth, turgor, water potential, and Young's modulus in pea internodes. *Physiologia Pl.* **20**, 213–231.

Cleland, R. E. 1967 A dual role of turgor pressure in auxin-induced cell elongation in *Avena* coleoptiles. *Planta* **77**, 182–191.

Cleland, R. E. & Rayle, D. L. 1977 Auxin, H$^+$-excretion and cell elongation. *Bot. Mag. Tokoyo, Spec. Issue* **1**, 125–139.

Cosgrove, D. J. 1981a Analysis of the dynamic and steady-state responses of growth rate and turgor pressure to changes in cell parameters. *Pl. Physiol.* **68**, 1439–1446.

Cosgrove, D. J. 1981b Rapid suppression of growth by blue light. Occurrence, time course, and general characteristics. *Pl. Physiol.* **67**, 584–590.

Cosgrove, D. J. 1982 Rapid inhibition of hypocotyl growth by blue light in *Sinapis alba* L. *Pl. Sci. Lett.* **25**, 305–312.

Cosgrove, D. J. & Cleland, R. E. 1983a Solutes in the free space of growing stem tissues. *Pl. Physiol.* **72** (in press).

Cosgrove, D. J. & Cleland, R. E. 1983b Osmotic properties of pea internodes in relation to growth and auxin action. *Pl. Physiol.* **72** (in press).

Cosgrove, D. J. & Green, P. B. 1981 Rapid suppression of growth by blue light. Biophysical mechanism of action. *Pl. Physiol.* **68**, 1447–1453.

Dainty, J. 1976 Water relations in plant cells. In *Encyclopedia of plant physiology* (ed. U. Luttge & M. G. Pitman), vol. 2A, pp. 12–35. New York: Springer-Verlag.

Falk, S., Hertz, C. & Virgin, H. 1958 On the relation between turgor pressure and tissue rigidity. I. Experiments on resonance frequency and tissue rigidity. *Physiologia Pl.* **11**, 802–817.

Firn, R. D. & Digby, J. 1980 The establishment of tropic curvatures in plants. *A. Rev. Pl. Physiol.* **31**, 131–148.

Franssen, J. M. & Bruinsma, J. 1981 Relationships between xanthoxin, phototropism, and elongation growth in the sunflower seedling *Helianthus annuus* L. *Planta* **151**, 365–370.

Gaba, V. & Black, M. 1979 Two separate photoreceptors control hypocotyl growth in green seedlings. *Nature, Lond.* **278**, 51–54.

Göring, H., Möller, H.-P. & Bleiss, W. 1975 Short-term kinetics of extension growth and Young's modulus of wheat coleoptiles caused by IAA-application and changes of water potential. *Biochem. Physiol. Pfl.* **168**, 411–420.

Green, P. B. & Cummins, W. R. 1974 Growth rate and turgor pressure. Auxin effect studied with an automated apparatus for single coleoptiles. *Pl. Physiol.* **54**, 863–870.

Green, P. B., Erickson, R. O. & Buggy, J. 1971 Metabolic and physical control of cell elongation rate – *in vivo* studies in *Nitella*. *Pl. Physiol.* **47**, 423–430.

Hall, A. B., Firn, R. D. & Digby, J. 1980 Auxins and shoot tropisms – a tenuous connection? *J. biol. Educ.* **14**, 195–199.

Holmes, M. G. & Schäfer, E. 1981 Action spectra for changes in the "high irradiance reaction" in hypocotyls of *Sinapis alba* L. *Planta* **153**, 267–272.

Hüsken, D., Steudle, E. & Zimmermann, U. 1978 Pressure probe technique for measuring water relations of cells in higher plants. *Pl. Physiol.* **61**, 158–161.

Iino, M. 1982 Inhibitory action of red light on the growth of maize mesocotyl: evaluation of the auxin hypothesis. *Planta* **156**, 388–395.

Kende, H. & Lang, A. 1964 Gibberellins and light inhibition of stem growth in peas. *Pl. Physiol.* **39**, 435–440.

Lockhart, J. A. 1965 An analysis of irreversible plant cell elongation. *J. theor. Biol.* **8**, 264–275.

Michelena, V. A. & Boyer, J. S. 1982 Complete turgor maintenance at low water potentials in the elongating region of maize leaves. *Pl. Physiol.* **69**, 1145–1149.

Meijer, G. 1968 Rapid growth inhibition of gherkin hypocotyls in blue light. *Acta bot. neerl.* **17**, 9–14.

Molz, F. J. & Boyer, J. S. 1978 Growth-induced water potentials in plant cells and tissues. *Pl. Physiol.* **62**, 423–429.

Molz, F. J. & Ferrier, J. M. 1982 Mathematic treatment of water movement in plant cells and tissue: a review. *Pl. Cell Envir.* **5**, 191–206.

Morgan, D. C., O'Brien, T. & Smith, H. 1980 Rapid photomodulation of stem extension in light-grown *Sinapis alba* L. Studies on kinetics, site of perception and photoreceptor. *Planta* **150**, 95–101.

Murray, P. B. & Acton, G. J. 1974 The role of gibberellin in hypocotyl extension of dark-growing *Lupinus albus* seedlings. *Planta* **117**, 209–217.

Nilsson, B., Hertz, C. & Falk, S. 1958 On the relation between turgor pressure and tissue rigidity. II. Theoretical calculations on model systems. *Physiologia Pl.* **11**, 818–837.

Pickard, B. G. & Thimann, K. V. 1964 Transport and distribution of auxin during tropistic response. *Pl. Physiol.* **39**, 341–350.

Ray, P. M. 1974 The biochemistry of the action of indoleacetic acid on plant growth. *Rec. Adv. Phytochem.* **7**, 93–122.

Ray, P. M. & Ruesink, A. W. 1962 Kinetic experiments on the nature of the growth mechanism in oat coleoptile cells. *Devl Biol.* **4**, 377–397.

Ray, R. M. & Ruesink, A. W. 1963 Osmotic behavior of oat coleptile tissue in relation to growth. *J. gen. Physiol.* **47**, 83–101.

Ross, C. L. & Rayle, D. L. 1982 Evaluation of H^+ secretion relative to zeatin-induced growth of detached cucumber cotyledons. *Pl. Physiol.* **70**, 1470–1474.

Russel, D. & Galston, A. 1969 Blockage by gibberellic acid of phytochrome effects on growth, auxin responses, and flavonoid synthesis in etiolated pea internodes. *Pl. Physiol.* **44**, 1211–1216.

Sherwin, J. E. & Furuya, M. 1973 A red far-red reversible effect on uptake of exogenous IAA in etiolated rice coleoptiles. *Pl. Physiol.* **51**, 295–298.

Stuart, D. A. & Jones, R. L. 1978 The role of acidification in gibberellic acid- and fusicoccin-induced elongation growth of lettuce hypocotyl sections. *Planta* **142**, 135–145.

Thomas, B. & Dickinson, H. G. 1979 Evidence for two photoreceptors controlling growth in de-etiolated seedlings. *Planta* **146**, 545–550.

Thornton, R. M. & Thimann, K. V. 1967 Transient effects of light on auxin transport in the *Avena* coleoptile. *Pl. Physiol.* **42**, 247–257.

Trewavas, A. 1981 How do plant growth substances work? *Pl. Cell Envir.* **4**, 203–228.

Uhrström, I. 1969 The time effect of auxin and calcium on growth and elastic modulus in hypocotyls. *Physiologia Pl.* **22**, 271–287.

Wainwright, S. A. 1970 Design in hydraulic organisms. *Naturwissenschaften* **57**, 321–326.

Walton, J. & Ray, P. M. 1981 Evidence for receptor function of auxin binding sites in maize. Red light inhibition of mesocotyl elongation and auxin binding. *Pl. Physiol.* **68**, 1334–1338.

Vanderhoef, L. & Briggs, W. R. 1978 Red light-inhibited mesocotyl elongation in maize seedlings. I. The auxin hypothesis. *Pl. Physiol.* **61**, 534–537.

Virgin, H. 1955 A new method for the determination of the turgor of plant tissues. *Physiologia Pl.* **8**, 954–962.

Zeiger, E., Field, C. & Mooney, H. 1981 Stomatal opening at dawn: possible roles for the blue light response in nature. In *Plants and the daylight spectrum* (ed. H. Smith), pp. 391–408. London: Academic Press.

Zimmermann, U. & Steudle, E. 1978 Physical aspects of water relations of plant cells. *Adv. bot. Res.* **6**, 45–117.

Discussion

A. W. GALSTON (*Plant Breeding Institute, Trumpington, U.K.*). Many years ago, Blaauw, in the Netherlands, investigated the 'light growth reaction' (*Lichtwachstumsreaktion*) in various tissues, including grass coleoptiles and *Phycomyces* sporangiophores. In etiolated coleoptiles, exposure to a pulse of blue light was followed by a temporary depression of growth rate, then an overshoot above the control dark value, and a return to the original growth rate. In *Phycomyces*, because of the sporangium lens effect, light induced a temporary rise in growth rate, followed by an undershoot, and a return to normal. After the completion of this entire cycle, there was no net effect of light on total growth in either organism. Dr Cosgrove's curves show no such details. Can he rationalize the differences between his results and Blaauw's?

D. COSGROVE. The figures I presented only show the *inhibition* of the growth rate, as that is the part of the response that we have studied most intensely. We have also followed the recovery of the high growth rate after a pulse of blue light; typically the recovery is slower than the inhibition, and there are pronounced oscillations in the growth rate (see our earlier published figures of the responses in pea, mustard, sunflower and cucumber seedlings). These oscillations frequently lead to an overshoot above the previous dark growth rate, but in general the overshoot is insufficient in magnitude and duration to cancel out completely the inhibitory effect of light. Such overshoots have been noted by other workers such as Gaba & Black.

Regarding Blaauw's results, I think the differences he has mentioned are due to the different organisms involved. It is well known that the *Phycomyces* growth response to a step-up in fluence rate is transient and is linked with the sporangiophore's phototropic behaviour. In dicot stems, in contrast, the growth inhibition persists as long as the stems are irradiated. Blaauw also examined the light growth response in sunflower hypocotyls, and found a pattern of response similar to ours.

Phil. Trans. R. Soc. Lond. B **303**, 467–478 (1983)

Printed in Great Britain

Photoperception and photomovement

By W. Haupt

Institute of Botany, University of Erlangen-Nürnberg, Schlossgarten 4, D-8520 Erlangen, F.R.G.

The photoperception systems in photomovement show a great diversity at several levels. (i) Different types of photomovement (namely photokinesis, photophobic response, phototaxis) make use of completely different perception systems even in the same organism. (ii) In different organisms a given type of photomovement is mediated by different and unrelated systems, as is shown in this paper particularly for phototaxis. (iii) In certain cases, a single response in one organism can be controlled by two separate and independent photoperception systems, which may even be located in different compartments.

This diversity sets photomovement apart from photosynthesis and vision, which are highly uniform and stable in evolution. Possible consequences for considering the evolution are discussed.

1. Introduction

For an autotrophic organism the most important environmental factor is light. Plants have therefore evolved what appear to be effective mechanisms to optimize the exploitation of the cosmic energy source, i.e. the sunlight. Among these mechanisms, different types of movement are known, which are controlled by light in different ways.

As a first example, unicellular or filamentous algae are mentioned, which move by gliding along solid surfaces in water. In these organisms the degrees of freedom of movement are restricted to two dimensions in space. Blue-green algae (Oscillatoriaceae, Nostocaceae) and diatoms have been well investigated in this respect, but some desmids and unicellular red algae also exhibit this type of movement.

A second example is given by flagellates, which are free-moving organisms: they can move in their aqueous medium in all dimensions of space.

In both of the above-mentioned types light can control or affect speed, continuity and direction of movement. The respective responses are termed photokinesis, photophobic response and phototaxis, as will be explained in the next section, and the general term is photomovement (Nultsch 1975; Häder 1979).

Under certain conditions these photomovements can lead to non-uniform distributions of the organisms in their environment, with patterns that are in some way related to the light. These results, macroscopically observed, are termed photoaccumulation, as long as no definite analysis has been made as to which of the photomovement types underlies the patterns (see Diehn 1979).

Besides the motility of whole organisms, the movement of chloroplasts in the cell can also be controlled by light. The resulting patterns of chloroplast distribution can be compared with photoaccumulation of motile organisms (see Britz 1979; Haupt 1982).

The following questions can be posed.

(i) If an organism exhibits different types of photomovement, is light perception identical in these different types?

(ii) If two organisms exhibit analogous photoresponses, do they use identical or at least comparable photoperception systems?

(iii) Are there organisms that make use of different photoperception systems for one response, independent from each other?

(iv) From the observations resulting from questions (i) to (iii), what are the conclusions about the evolution of photoperception?

2. LIGHT PERCEPTION IN DIFFERENT TYPES OF PHOTOMOVEMENT IN ONE ORGANISM

As an example, a well investigated genus of cyanobacteria (blue-green algae), *Phormidium*, will be described with its responses (see Häder 1979; Nultsch & Häder 1979). The organism consists of many cells in one-dimensional order, forming a trichome. The gliding movement requires contact with a solid surface and proceeds in the direction of the axis of the trichome. Since there is not an apico-basal polarity, both tips of the trichome are equivalent and can act as the advancing end or as the back end.

The different photoresponses are characterized as follows.

(i) Photokinesis: the steady-state velocity of movement depends on the intensity of constant light. It usually increases with increasing light, up to an optimum beyond which it decreases again.

(ii) Photophobic response: a sudden step down in light intensity evokes a transient response. The trichome ceases moving, and after a while it resumes movement in the opposite direction, soon reaching the steady-state velocity as dictated by photokinesis. Under certain conditions, a photophobic response can be induced also by a step up in light intensity. Thus the photophobic response interrupts the continuity of movement. However, even without a photophobic stimulus a reversal of movement can be found, but this autonomous reversal occurs with low frequency, i.e. every few minutes. Photophobic response, then, is a premature reversal.

(iii) Phototaxis: in unilateral light most of the organisms accumulate as far towards the light source as possible. Thus movement obviously has a relation to the light direction. This result is achieved by particular responses of the individual trichomes: if a trichome happens to glide in the 'wrong' direction (i.e. light coming from behind), it is induced to a premature reversal. If, however, movement proceeds in the 'right' direction (i.e. light coming from front), autonomous reversal is delayed. On average, therefore, movement towards the light prevails.

Thus the organism has to measure the steady-state light intensity for photokinesis, the constancy or change of intensity in time for photophobic response, and the light direction for phototaxis. In the next subsections, current knowledge about these photoperception processes in *Ph. uncinatum* will be described.

(a) *Photoperception in photokinesis*

In *Phormidium uncinatum* the action spectrum of photokinesis has maxima in the blue and red region, with peaks in the visible light around around 430 and 670 nm, and extending into the far red (figure 1*a*). This points to chlorophyll as the photoreceptor pigment.

Since absorption by the biliproteins contributes only slightly to the effect, photosystem I of photosynthesis is assumed to be important for photokinesis, and acceleration of movement with increasing intensity should simply be due to additional ATP from cyclic photophosphorylation.

Accordingly, uncouplers of photophosphorylation, e.g. desaspidin or carbonylcyanide *m*-chlorophenylhydrazone (CCCP), abolish the photokinetic effect. On the other hand, 3-(3,4-dichlorophenyl)-1,1-dimethylurea (DCMU), an inhibitor of the non-cyclic electron transport chain, has relatively little effect, corresponding to the ineffectiveness of the photosystem II pigments (Nultsch 1975; Häder 1979).

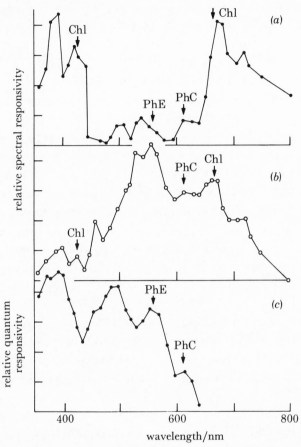

FIGURE 1. Action spectra for (*a*) photokinesis, (*b*) photophobic response and (*c*) phototaxis in *Phormidium uncinatum*. The main absorption peaks of chlorophyll *a* (Chl), phycoerythrin (PhE), and phycocyanin (PhC) are indicated. (Modified after Nultsch (1962).)

In conclusion, in photokinesis light is used as an additional energy source for movement. Thus, strictly speaking, light does not act as a signal, in contrast to the responses in the following subsections, which are energetically independent of the light stimulus.

(*b*) Photoperception in photophobic response

The step-down photophobic response is most effectively induced by light between 530 and 680 nm, with peaks close to the absorption maxima of C-phycoerythrin, C-phycocyanin and chlorophyll, whereas absorption in carotenoids and in the blue peak of chlorophyll contributes little to the response (figure 1*b*). This action spectrum is taken as evidence for photosystem II as the most important photoperception system (Nultsch 1962; Häder 1979).

Indeed, inhibitors of photosynthetic electron flow inhibit the photophobic response specifically, whereas uncouplers of photophosphorylation have litte effect (Häder 1979). In such experiments, photokinesis can be used as a control, because there the effects are opposite.

From these results it has been assumed that a sudden decrease in light intensity transiently depletes an electron pool in the electron transport chain by reducing the input from photosystem II, while the outflow to photosystem I is less influenced. This hypothesis leads to an interesting consequence: the same effect should be obtained if the electron input is kept constant but the pool is drained by an increased photosystem I activity. And indeed, a step-up signal, too, can be followed by a photophobic response, provided that this light is predominantly absorbed by photosystem I. In fact the mixture of effects makes things rather complicated, but clear conclusions have been obtained by stimulation with step-up and step-down signals of monochromatic light on the background of constant light of different wavelengths (Nultsch, 1975; Häder, 1979). Finally, information about the localization of the pool has been obtained by the application of redox substances. Depending on their redox potential, they feed in or abstract electrons at different sites of the electron transport chain. From comparative inhibitions of photophobic response by these substances, plastoquinone has been decided on as the most promising candidate for the electron pool involved in photophobic response (Häder 1979).

Although we have no full understanding of the subsequent steps, bioelectric effects of the changed redox state of plastoquinone (Häder & Poff 1982) are assumed to be involved, and indeed changes in cell potential have been measured as an early result of photophobic stimulus (Häder 1978). Moreover, the final response, i.e. reversal of movement, can be inhibited by applying an external electrical field in the direction of movement (Häder 1979). In conclusion, the photoperception processes are completely different in photokinesis and photophobic response.

(c) Photoperception in phototaxis

The phototactic action spectrum extends from *ca.* 350 nm or below to *ca.* 600 nm (figure 1 *c*). The peaks in the green and yellow region suggest that biliproteins act as photoreceptor pigments, but no firm conclusion is possible in the short-wavelength region, the peak near 490 nm suggesting carotenoids, but the near-ultraviolet peak pointing to a flavin (Nultsch 1962, 1975; Häder 1979).

Because chlorophyll is not involved, the perception system in phototaxis obviously has no relation to photosynthesis. This negative conclusion is almost the only knowledge we have about photoperception. In particular we do not yet understand the perception of directionality in *Phormidium*, i.e. the mechanism by which the organism can distinguish between light coming from the front or from the rear (see Haupt 1974).

As a result, the perception system of phototaxis in *Phormidium* is different from those of the two other photomovement responses. Particularly interesting is the comparison with the photophobic response: in both phototaxis and photophobic response the final effect is reversal of movement. But this reversal is induced in two completely different ways: in photophobic response by a temporal change in light intensity, via photosynthetic electron flow; in phototaxis by light coming continuously from the wrong direction, via an unknown chain of events. Thus the two photoperception systems differ fundamentally from each other, at least in their photoreceptor pigments and the parameters of the stimulus measured. Yet the two perception systems converge on a final response.

3. Uniformity or diversity in photoperception systems between different organisms?

Now the question can be asked whether the results in *Phormidium* can be generalized: that photokinesis and the photophobic response are mediated by photosynthesis, although via different mechanisms, but that phototaxis is mediated by a different system with blue light having the predominant effect. This question will be discussed for photokinesis and for phototaxis.

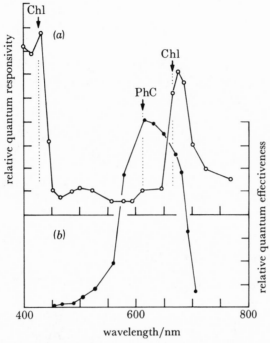

FIGURE 2. Action spectra for photokinesis in (*a*, ○) *Phormidium autumnale* and (*b*, ●) *Anabaena variabilis*. The main absorption peaks of chlorophyll *a* (Chl) and phycocyanin (PhC) are indicated by the arrows and dotted lines. (Modified after Nultsch (1975).)

(*a*) Photokinesis in different organisms

The cyanobacterium (blue-green alga) *Anabaena variabilis* exhibits photokinesis similar to that in *Phormidium*, but the action spectrum points to C-phycocyanin as the main photoreceptor pigment (figure 2). It has therefore been concluded, and confirmed by inhibitor studies, that in *Anabaena* photosystem II is important and that pseudocyclic photophosphorylation is the checkpoint for the light action on motility. This organism thus differs from *Phormidium*, where mainly photosystem I and cyclic photophosphorylation are involved. Other blue-green algae appear to be 'mixed types', making use of both photoperception systems (Nultsch 1975).

There are organisms, belonging to a diversity of groups, e.g. flagellates, diatoms and unicellular red algae, that exhibit photokinesis as well. In most of these cases, photosystem I or II, or both, of photosynthesis are involved, as in cyanobacteria, but in flagellates of the order Volvocales a pure blue-light effect is reported (Nultsch 1975; Häder 1979). It is unlikely that in that case light acts by providing additional energy for locomotion. Thus photokinesis can make use of different photoperception systems.

(b) Phototaxis in different organisms

An enormous diversity of perception systems in phototaxis becomes obvious if we compare phototactic action spectra of several organisms (figures 3 and 4). In a first group only blue light is effective. Several green flagellates like *Chlamydomonas*, *Platymonas* and *Volvox* (figure 3 *a*) have their maximum near 500 nm, and this is usually interpreted as a carotenoid or carotenoprotein as photoreceptor pigment (see, for example, Ferrara 1975; Schletz 1976;

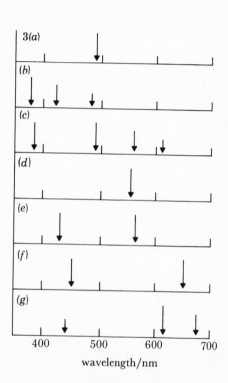
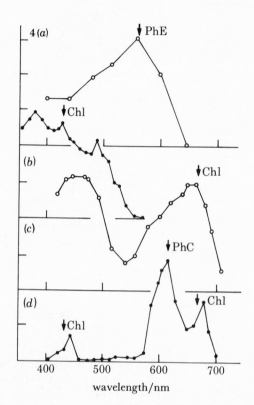

FIGURE 3. Spectral distribution of photoactic activity in different organisms: (*a*) *Volvox aureus*, (*b*) *Nitzschia communis*, (*c*) *Phormidium autumnale*, (*d*) *Cryptomonas* sp., (*e*) *Dictyostelium discoideum*, (*f*) *Micrasterias denticulata* and (*g*) *Anabaena variabilis*. The arrows indicate peaks in the respective action spectra, with their length being a qualitative measure of the peak height (for *d*, *b*, *f* and *g* compare the action spectra in figure 4). (After data of Neuscheler (1967), Nultsch (1975), Poff *et al.* (1973) and Watanabe & Furuya (1974).)

FIGURE 4. Action spectra for phototaxis: (*a*) *Cryptomonas* sp.; (*b*) *Nitzschia communis*; (*c*) *Micrasterias denticulata*; (*d*) *Anabaena variabilis*. The ordinate denotes quantum responsivity (*a*, *b*, *d*) or energy effectiveness (*c*). The main absorption peaks of chlorophyll *a* (Chl), phycoerythrin (PhE) and phycocyanin (PhC) are indicated. (Modified after Neuscheler (1967), Nultsch (1975) and Watanabe & Furuya (1974).)

Feinleib 1978; Nultsch & Häder 1979), but a rhodopsin-like pigment is also under discussion (Foster & Smyth 1980). More difficult to interpret is the compound action spectrum of the diatom *Nitzschia communis* (figures 3*b* and 4*b*). On the one hand, the strong peak in the near u.v. points to a flavin, but the blue peak is shifted to longer wavelengths, which appears more consistent with carotenoids; the latter view is supported by the observation that the colourless, carotenoid-free *Nitzschia alba* is non-phototactic (Nultsch & Häder 1979).

In a second group of organisms, light between 500 and 600 nm, i.e. green and yellow light, is highly effective in phototaxis. As has been shown already, phycoerythrin and phycocyanin contribute to photoperception in *Phormidium*, besides the short-wavelength light (figure 3*c*; see also figure 1), and in the flagellate *Cryptomonas*, phycoerythrin appears to be the only photoreceptor pigment (figures 3*d* and 4*a*) (Watanabe & Furuya 1974). A similar spectral sensitivity for phototaxis in green light has been reported in the slime mould *Dictyostelium* (figure 3*e*), which, however, does not contain phycoerythrin; instead, a haem compound is proposed for photoreception (Poff *et al.* 1973; Poff & Butler 1974).

Finally, there are also examples with pronounced red-light effects, which point to an important role of the photosynthetic pigments in photoperception. In the desmid *Micrasterias*, the action spectrum peaks in the blue and in the red with almost equal effectiveness (figures 3*f* and 4*c*) (Neuscheler 1967), and in the cyanobacterium *Anabaena variabilis* the maximum coincides with the absorption of the accessory pigment of photosynthesis, phycocyanin, besides two less effective chlorophyll peaks in the blue and the red (figures 3*g* and 4*d*) (Nultsch 1975). Thus the photosynthetic pigments obviously act as photoreceptors for phototaxis in desmids and in *Anabaena* (see also Häder 1981), but in most of the other examples investigated no relation seems to exist between phototaxis and photosynthesis. Moreover, within this latter group a great variety of possibilities is suggested, as concluded from the action spectra (see figure 3*a–e*).

This diversity of photoreceptor pigments in phototaxis can be considered in the context of the diversity of orientation mechanisms. As has been stated above, in *Phormidium* the response simply consists of a modification of autonomous reversal of the gliding direction, induced by the direction of light. A similar mechanism has been found in some diatoms (see Nultsch 1975). Flagellates, on the other hand, orient by active steering, but it is not yet definitely known whether this steering is performed in the same way in all cases. Only in *Euglena* has the behaviour been analysed with some certainty. Here the cell rotates during movement; if it is not aligned with the light path, it receives a signal once each revolution, according to which it corrects its trail (Diehn 1979). Similarly the steering in *Volvox* is connected with the rotation of the colony, by which each individual cell experiences a temporal pattern of light absorption in unilateral illumination (Schletz 1976).

The trichomes of *Anabaena* steer towards the light as well, but in this case no rotation is involved and hence the directional signal has to be perceived continuously. Accordingly a continuous curving of the gliding trichome into the light path is observed as long as it moves (Nultsch 1975).

Finally, the gliding desmid *Micrasterias* steers continuously as well, but an additional peculiarity has been found in this organism (Neuscheler 1967). If a stationary cell is illuminated from a direction other than the direction of potential gliding, the cell rotates to the correct orientation, ready to glide straight towards the light as soon as movement is resumed.

As a result, much more diversity in photoperception systems exist in phototaxis than in photokinesis, and this diversity concerns both the input and the output of these systems, i.e. the photoreceptor pigments and the mechanisms of orientation, respectively. In detail, nearly all problems of the transduction processes are still open questions.

4. RESPONSES MEDIATED BY MORE THAN ONE PHOTOPERCEPTION SYSTEM

In the preceding sections some examples have been presented as evidence for different photoreceptor pigments underlying the same response in one organism: phototaxis in *Anabaena* uses chlorophyll *a* and C-phycocyanin (§3*b* and figure 4*d*). Likewise, for the photophobic response in *Phormidium*, light absorption in chlorophyll *a* as well as in phycoerythrin and phycocyanin is important (§2*b* and figure 1*b*). In these cases, however, the two or three pigments feed the light energy into photosynthesis, and thus only the very first step of photoperception is duplicated.

Less well understood is the example of phototaxis in *Phormidium* (§2*c* and figure 1*c*), where obviously light absorption by a yellow pigment and that by the biliproteins can start the photoperception process. We do not yet know where these two origins converge to a common chain of processes.

There is, however, one example that clearly demonstrates completely independent photo-perception systems for the same response in one organism. This is light-oriented chloroplast displacement in the green alga *Hormidium*. The cylindrical cells of *Hormidium* contain one large chloroplast only, which covers about half the cell circumference (figure 5*a*). It can slide along

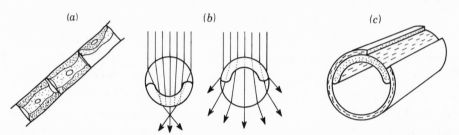

FIGURE 5. Chloroplast movement in *Hormidium flaccidum*. (*a*) Part of a trichome, showing the chloroplast facing the front or rear wall in two cells and facing the flank wall in one cell. (*b*) Cross sections of a cell in air (left) or in oil (right), showing refraction of parallel light rays. The resulting absorption gradient orients the chloroplast to the rear (left) or to the front (right). (*c*) Schematic drawing of a cell with part of the cytoplasm removed (cell wall omitted). The dashes represent the main transition moments of the photoreceptor molecules, with longitudinal orientation in the chloroplast and transverse orientation in the cytoplasm. (After Scholz 1976*b*), Haupt (1983) and Haupt & Wagner (1983).)

the cell wall, thus orienting to light. In detail, this orientation depends not only on the light intensity but also on the surrounding medium: in low-intensity white light (less than 40 W m^{-2}), the chloroplast is preferentially found distal to the light source if the cell is surrounded by air, but proximal if paraffin oil is the surrounding medium. In the former case, the cell acts as a collecting lens, in the latter case as a diverging lens, due to the differences in the refractive indices (figure 5*b*). Thus the chloroplast always prefers the brightest region in the peripheral cytoplasm (Scholz 1976*a*).

These responses are restricted to the blue-light region of the spectrum, and they exhibit an action dichroism, with polarized light being more effective if its electrical vector vibrates perpendicular rather than parallel to the cell axis. Potassium iodide, which is known as a quenching substance of the triplet excited state of flavins, inhibits the response; and a similar effect is found if the flavin content of the cell is reduced by a dark treatment for a couple of days. Finally, responsivity is resumed if iodide is washed out, or if the flavin-deficient cells are externally provided with flavin mononucleotide (FMN). There is therefore very good evidence for a flavin's being the photoreceptor pigment, and for its localization in the peripheral cytoplasm with a surface-parallel orientation of the transition moments, preferentially in a transverse direction (figure 5*c*).

So far, this description is a simplification, as detailed data show. In fact, the treatments with KI or with darkness do not abolish the response completely, but the chloroplast still orients.

However, under these conditions there is an increased tendency to approach the proximal region in all surrounding media, i.e. to orient towards the light source (figure 6). Obviously the lens effect becomes less important, and we have to assume another perception system that enables the cell to measure the light direction independently of light refraction. Information about the second photoperception system is obtained by additional observations: it operates at a broader range of fluence rates than the first system; the action spectrum has an accessory peak in red light, and this red-light effect is sensitive to DCMU; finally, the action dichroism is reversed, parallel-vibrating blue light being the most effective. These facts led to the assumption that this second system is localized in the chloroplast rather than in the cytoplasm (figure 5c), and that it has the photosynthetic pigments as photoreceptors. The light direction is sensed as a result of attenuation as light traverses the chloroplast, thus establishing a gradient of light intensity through the chloroplast (Scholz 1976b; see also Haupt 1983).

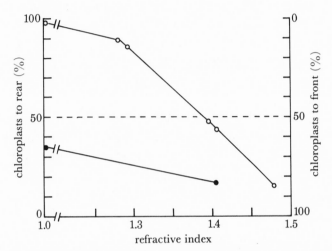

FIGURE 6. Chloroplast orientation in *Hormidium* to the front or the rear, respectively, as dependent on the refractive index of the environment (cf. figure 5b), and on the pretreatment (daily light-dark cycles (○) or 5 days of continuous darkness (●)). The broken line indicates a random distribution to front and rear positions. (After Haupt (1983), from data of Scholz (1976b).)

In conclusion, in *Hormidium* we have two clearly distinguishable photoreceptor pigments, located in two different compartments of the cell, and this also seems to be true in higher plants (Seitz 1979). Light absorption, then, starts two separate and independent perception processes, which finally have the same result: movement of the chloroplast in the direction of higher light absorption. Remarkably, this same result is found whether an absorption gradient is established in the moving organelle, i.e. in the chloroplast, or in its resting environment, i.e. in the cytoplasm at the membrane. It can be assumed that both perception processes start an elementary biochemical or biophysical process at some point, e.g. a change in active or passive ion fluxes across a membrane, as discovered in the alga *Vaucheria* (Blatt *et al.* 1981), and there is good evidence that those changes could activate the motor apparatus for chloroplast displacement, probably actin microfilaments in the cytoplasm (see Britz 1979; Haupt & Wagner 1983). This could be compared, then, with the activation of animal or human muscles by different types of stimulus, perceived by sensory systems that differ in structure, function and localization.

5. Evolutionary aspects of photomovement

(a) Selection of photoperception systems in evolution

It has been shown that in photomovement a great diversity exists concerning photoperception. This diversity is twofold: it concerns different types of response in one organism (§2), and it concerns one response among different organisms (§3). Thus photomovement differs fundamentally from photosynthesis, where a common photoperception system can be found throughout almost the whole plant kingdom. It also differs from vision in animals, where rhodopsin is found as the photoreceptor pigment throughout, irrespective of very different cytological structures of the sensory systems, and irrespective of differences in details of the secondary processes.

This suggests differences in the principle of evolution between the uniform and pluralistic systems, i.e. photosynthesis and vision on the one hand, and photomovement on the other hand. In the former cases, the pressure of selection must have been very heavy, allowing only one optimal system in each to survive. In the latter case, however, several trials of evolution have survived, which very probably are not completely equivalent to each other in their efficiency and in their economy.

This allows speculation that the photoreceptor pigments for photomovement responses have not evolved primarily for these responses. Instead, only an additional use of these already existing pigments had to be invented and to be established, together with completing the corresponding transduction chains (i.e. the perception processes in the sense of the present symposium).

Consistent with this speculation is especially the fact that two different perception chains for one response can sometimes be found in one organism, as has been reported for the chloroplast orientation in *Hormidium* (§4). Evolution of those cases is hard to understand solely on the basis of selection: as soon as one system has evolved, there are no reason and no means for a second system to be selected. Here the only reasonable explanation seems to be the pre-existence of parts of the perception systems, which must have been evolved for other reasons.

(b) Ecology of photomovement

If this speculation is correct, photomovement should prove to be advantageous for the organism, but not an absolute necessity for survival. I shall therefore try to come to a conclusion about the biological significance of photomovement.

The significance of *photokinesis* is not immediately obvious. Usually photokinesis is positive, i.e. the organism moves faster in light than in darkness. For an autotrophic organism one could speculate that this faster movement during periods of high photosynthesis would facilitate uptake of nutritional salts, which may be limiting: the organism could exploit a larger volume of the medium without exhausting it. However, such an interpretation, although plausible, does not take into consideration the density of organisms in the natural habitat, nor turbulences in the medium. Hence the interpretation is probably weak.

Much more obvious is the significance of the *photophobic response*, which is typically a step-down response. In an aqueous habitat that is partly shaded, e.g. by leaves, organisms live in a pattern of bright and dim regions. If an organism, during its movement, happens to cross a border from bright to dim, a photophobic response often brings it back to the bright region, thus avoiding an unfavourable reduction in photosynthesis. The bright regions, then, can act as light traps.

On the first view, *phototaxis* appears to be a very useful response: it brings organisms to the optimal light conditions by an active orientation. Indeed, this is true in the laboratory, where light intensity usually increases as the organism approaches the light source. In Nature, this situation can be found in partly shaded habitats: organisms (and non-living particles) in the lighted region are sources of effective scattering of light falling directly on them. Thus, organisms in the shaded region 'see' the scattered light, they respond phototactically, they orient to and they finally accumulate in the bright regions, where, by the way, they are trapped by their photophobic response. If, however, an organism were to move phototactically towards direct sunlight it would not experience any increase in light intensity; nor would the intensity decrease if the organism moved straight away from direct sunshine. Thus the usual interpretation has to be reconsidered, according to which negative phototaxis can be a means to protect the organism against damaging irradiation.

6. GENERAL CONCLUSION

In research on photosynthesis and vision, much effort has been made to understanding the processes connecting light absorption with the final result. There is good reason to expect that many details of these findings can be generalized, at least within large groups of organisms. In photomovement, the situation is completely different. Even if, in a model organism, one perception system were analysed to the last detail, including all processes through to the final response, nothing could be concluded for other photomovement systems in that organism. The only generalization that can be made for photomovement is its diversity.

REFERENCES

Blatt, M. R., Weisenseel, M. H. & Haupt, W. 1981 A light-dependent current associated with chloroplast aggregation in the alga *Vaucheria sessilis*. *Planta* **152**, 513–526.

Britz, S. J. 1979 Chloroplast and nuclear migration. In *Encyclopedia of plant physiology* (new series), vol. 7 (ed. W. Haupt & M. E. Feinleib), pp. 170–205. Berlin, Heidelberg and New York: Springer-Verlag.

Diehn, B. 1979 Photic responses and photosensory transduction in motile protists. In *Handbook of sensory physiology*, vol. VII/6A (ed. H. Autrum), pp. 24–78. Berlin, Heidelberg and New York: Springer-Verlag.

Feinleib, M. E. 1978 Photomovement of microorganisms. *Photochem. Photobiol.* **27**, 849–854.

Ferrara, R. 1975 General review on phototactic action spectra. In *Biophysics of photoreceptors and photobehaviour of microorganisms* (ed. G. Colombetti), pp. 121–145. Pisa: Lito Felici.

Foster, K. W. & Smyth, R. D. 1980 Light antennas in phototactic algae. *Microbiol. Rev.* **44**, 572–630.

Häder, D.-P. 1978 Extracellular and intracellular determination of light-induced potential changes during photophobic reactions in blue-green algae. *Arch. Microbiol.* **119**, 75–79.

Häder, D.-P. 1979 Photomovement. In *Encyclopedia of plant physiology* (new series), vol. 7 (ed. W. Haupt & M. E. Feinleib), pp. 268–309. Berlin, Heidelberg and New York: Springer-Verlag.

Häder, D.-P. 1981 Effects of inhibitors on photomovement in desmids. *Arch. Microbiol.* **129**, 168–172.

Häder, D.-P. & Poff, K. L. 1982 Spectrophotometric measurement of plastoquinone photoreduction in the blue-green alga, *Phormidium uncinatum*. *Arch. Microbiol.* **131**, 347–350.

Haupt, W. 1974 Phototactic movements in plants. In *Progress in photobiology* (*Proc. VI Int. Congr. Photobiol.*) (ed. G. O. Schenk), p. 26. Frankfurt: Deutsche Gesellschaft für Lichtforschung.

Haupt, W. 1982 Light-mediated movement of chloroplasts. *A. Rev. Pl. Physiol.* **33**, 205–233.

Haupt, W. 1983 Movement of chloroplasts under the control of light. In *Progress in phycological research 2* (ed. F. E. Round & D. J. Chapman), pp. 227–281. Amsterdam: Elsevier Science Publishers.

Haupt, W. & Wagner, G. 1983 Chloroplast movement. In *Membranes and sensory transduction* (ed. G. Colombetti & F. Lenci). London: Plenum Press. (In the press.)

Neuscheler, W. 1967 Bewegung und Orientierung bei *Micrasterias denticulata* Bréb. im Licht. *Z. PflPhysiol.* **57**, 151–172.

Nultsch, W. 1962 Phototaktische Aktionsspektren von Cyanophyceen. *Ber. dt. Bot. Ges.* **75**, 443–453.

Nultsch, W. 1975 Phototaxis and photokinesis. In *Primitive sensory and communication systems* (ed. M. J. Carlile), pp. 29–90. London, New York and San Francisco: Academic Press.

Nultsch, W. & Häder, D.-P. 1979 Photomovement of motile microorganisms. *Photochem. Photobiol.* **29**, 423–437.

Poff, K. L. & Butler, W. L. 1974 Spectral characteristics of the photoreceptor pigment of phototaxis in *Dictyostelium discoideum. Photochem. Photobiol.* **20**, 241–244.

Poff, K. L., Butler, W. L. & Loomis, W. F. Jr 1973 Light-induced absorbance changes associated with phototaxis in *Dictyostelium. Proc. natn. Acad. Sci. U.S.A.* **70**, 813–816.

Schletz, K. 1976 Phototaxis bei *Volvox*-Pigmentsysteme der Lichtrichtungsperzeption. *Z. PflPhysiol.* **77**, 189–211.

Scholz, A. 1976a Lichtorientierte Chloroplastenbewegung bei *Hormidium flaccidum*: Perception der Lichtrichtung mittels Sammellinseneffekt. *Z. PflPhysiol.* **77**, 406–421.

Scholz, A. 1976b Lichtorientierte Chloroplastenbewegung bei *Hormidium flaccidum*: Verschiedene Methoden der Lichtrichtungsperception und die wirksamen Pigmente. *Z. PflPhysiol.* **77**, 422–436.

Seitz, K. 1979 Cytoplasmic streaming and cyclosis of chloroplasts. In *Encyclopedia of plant physiology* (new series), vol. 7 (ed. W. Haupt & M. E. Feinleib), pp. 150–169. Berlin, Heidelberg and New York: Springer-Verlag.

Watanabe, M. & Furuya, M. 1974 Action spectrum of phototaxis in a cryptomonad alga, *Cryptomonas* sp. *Pl. Cell Physiol.* **15**, 413–420.

Discussion

A. W. GALSTON (*Plant Breeding Institute, Cambridge, U.K.*). (1) Professor Haupt showed a slide on which a filament of *Anabaena* had made a right-angled turn in the direction of a unilateral light source. Is anything known about the physical basis for such a movement? (2) What is known about the mechanism of the gliding movements of *Phormidium*?

W. HAUPT. (1) Nothing is known about the mechanism of gliding in *Anabaena*, even if it moves straight forward. Hence it is premature to ask for the mechanism of steering. Theoretically, for an orientation to take place it must be assumed that a gradient of light absorption in a cell results in different light effects on the motor apparatus at the lighted and shaded sides.

(2) In *Phormidium* the trichome moves while rotating around its axis. Phenomenologically this is identical with the gliding in *Oscillatoria*, and hence extrapolation from this organism may be justified. Here, undulations of an outer layer of the cell wall appear to be the cause of gliding, including rotation, and the undulations are supposed to be brought about by the activity of fibrillar structures found beneath this layer. These structures are probably different from actin microfilaments, because we are dealing with prokaryotic organisms, which do not possess actin.

J. BENNETT (*Department of Biological Sciences, University of Warwick, U.K.*). Which phytochrome-mediated responses seem to involve the fewest steps between photoreception and action?

G. WAGNER (*Botanishces Institut der Justus-Liebig-Universität, Giessen, F.R.G.*). So far there is no case where the complete series of steps between photoreception and action has been analysed. Those cases in photomorphogenesis where regulation of gene expression is involved, probably contain many steps. On the other hand, relatively few steps are assumed for phytochrome-mediated chloroplast orientation in *Mougeotia*. In this regard, I refer to the posters [exhibited outside the lecture hall during the meeting] by Haupt *et al.* and in particular by Wagner *et al.* entitled 'Calcium vesicles and calmodulin from *Mougeotia*'. The main topics of this poster are: isolation and characterization of calcium vesicles from *Mougeotia*; isolation and identification of calmodulin from *Mougeotia*; location of actin within *Mougeotia* by fluorescent phallotoxin. These data, taken together with earlier findings [shown in the poster by Haupt *et al.*], indicate a phytochrome action in *Mougeotia* via calcium–calmodulin–actomyosin to perform light-induced chloroplast reorientation in *Mougeotia*.

Phil. Trans. R. Soc. Lond. B **303**, 479–487 (1983)
Printed in Great Britain

Perception of a unilateral light stimulus

By K. L. Poff

Michigan State University – Department of Energy, Plant Research Laboratory,
Michigan State University, East Lansing, Michigan 48824, U.S.A.

An organism can detect light direction given a gradient in light intensity within the organism. This gradient, which may be measured temporally or spatially, can be produced by screening or by refraction. The ramifications of the method of producing the gradient are potentially great, with possible effects on the shape of dose–response curves and action spectra. Two biological systems, amoebal phototaxis in *Dictyostelium* and phototropism by monocot seedlings, illustrate some potential problems. In the former system, no obvious mechanism exists for producing a substantial internal gradient in light intensity. This indicates our lack of knowledge concerning the amount of gradient necessary for an organism to measure light direction. In the latter system, it is evident that a gradient in light intensity is established by screening for second positive phototropism. However, screening may not be the method used for first positive phototropism. The implications of refraction as the mechanism involved in first positive phototropism are sufficiently great to warrant a thorough examination of the role of screening and refraction in first positive phototropism.

Introduction

Many examples are known in which an organism perceives not only light but also the direction in which the light is propagated. Typically, this results in a movement or growth response related in some way to the directional stimulus. (This discussion will for ease of expression use the term 'light direction' to refer to the direction in which light is propagated.) For an organism to detect light direction, there must be some difference or gradient in light intensity within the organism, which can be translated into a gradient in light absorption. The gradient in light intensity can be measured in space (a spatial measurement) or in time (a temporal measurement), but it is the gradient which permits the detection of light direction regardless of the basis for its measurement (whether spatial or temporal).

This discussion will present the known methods whereby an internal gradient in light intensity can be produced and some of the ramifications that derive from these methods. Finally, two biological examples will be described to illustrate some of the more intriguing problems.

Screening

The mechanisms available for establishing an internal light gradient are screening and refraction (figure 1). (A dichroic receptor pigment may be used to measure the plane in which the light is propagated, but in the absence of screening or refraction, a dichroic receptor pigment cannot be used to measure light direction.) Screening decreases the light intensity beyond the screen relative to that before the screen, and may occur as a result of scattering or absorption. Scattering must be assumed to occur in all organisms, although the extent of the scattering may vary. Thus, no biological system should exist in which screening is solely by absorption. However, by definition, a photoresponsive system must contain a photoreceptor pigment, which must contribute some absorption to the tissue. It is therefore not theoretically possible to have an organism in which the screening is solely by scattering, and scattering and absorption both must contribute to screening, although either may be relatively insignificant.

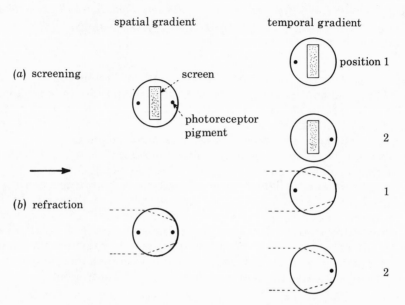

FIGURE 1. Schematic drawing of the mechanisms by which a gradient in light intensity can be established across an organism illuminated from the left. The drawings on the left depict a spatial perception of the gradient and the drawings on the right depict a temporal perception of the gradient. For a spatial perception, with a gradient established either by screening or refraction, a minimum of two detections must occur at two places in the gradient. For a temporal perception, two detections must also occur but separated in time rather than space. This separation in time is represented by the organism's moving from position 1 to position 2. (Adapted from Feinleib (1980).)

If the screening is largely by absorption, then it can be reasoned that the response is dependent upon absorption of light by the photoreceptor pigment and, in addition, is dependent upon absorption of light by the screening pigment (Thimann & Curry 1961). It follows directly from this argument that one of the major constraints for action spectroscopy is difficult to meet with a system in which the light intensity gradient is established by absorption. Namely, the response or action is not dependent only on the photoreceptor pigment but is dependent on two pigments. An action spectrum measured for such a system will indicate a complex product of the absorption spectrum of the photoreceptor pigment and the absorption spectrum of the screening pigment. Similarly, if the photoreceptor pigment itself serves as the screening pigment, the action spectrum will represent a complex product of the absorption spectrum of the photoreceptor pigment multiplied by itself.

REFRACTION

Refraction of light at a curved air–organism interface can focus the light within the organism. For an organism with a circular cross section and a relatively low internal absorbance, this results in a higher light intensity and a longer pathlength over which light can be absorbed on the distal side than on the proximal side (figure 2). Given the assumptions that the photoreceptor pigment is evenly distributed throughout the organism and that the initial reaction products do not readily diffuse throughout the organism, a greater number of pigment molecules will be 'excited' on the distal side than on the proximal side and thus the light direction will be detected. In a positive phototaxis or phototropism, movement or growth would be away from the side with the greater number of 'excited' pigment molecules.

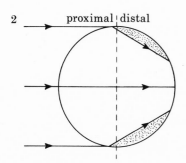
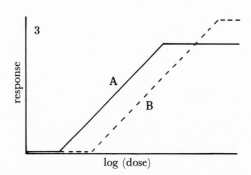

FIGURE 2. Schematic drawing of the cross section of an organism exposed to unilateral light. The lines represent light rays from a light source to the left of the organism. Because of refraction at the air–organism interface, light is focused onto the distal side of the organism. Note the darkened areas on the distal side of the organism.

FIGURE 3. Hypothetical dose–response curves for an organism that detects light direction by using a refraction-generated light gradient. Curve A represents the photoproduct formation per unit volume on the distal side; curve B represents photoproduct formation per unit volume on the proximal side. It is assumed that the photoreceptor pigment is evenly distributed and that the primary photoproducts are not readily diffusible throughout the organism.

One might expect a hypothetical stimulus response curve like figure 3a, with response increasing with the logarithm of the stimulus to saturation for the distal side of the organism. This saturation can result from any rate-limiting reaction in the stimulus–response sequence. Because the receptor pigment stimulation is greater on the distal side, saturation would occur first on that side. At a still higher fluence rate, saturation also would occur on the proximal side (figure 3b). However, because of the lens effect, only a portion of the distal side will be illuminated whereas almost the entire proximal side will be illuminated. This would result in a greater number of pigment molecules being stimulated on the proximal side than on the distal side. Thus, if one considers the number of photoproducts found on the two sides, saturation on the proximal side will occur not only at a higher fluence rate but also at a higher 'response' level than on the distal side. At these higher fluence rates, if growth or movement were still away from the side with the greater number of 'excited' pigment molecules, then the organism would grow or move away from the light.

Saturation might be expected at lower doses with a high fluence rate than with a low fluence rate. In such a case, the extent of a 'negative' response would be fluence-dependent.

BIOLOGICAL EXAMPLES

This discussion will not attempt to review the many biological systems where evidence is available concerning the mechanism whereby a light gradient is established and light direction detected. Rather, two biological examples will be discussed to illustrate some of the areas of uncertainty.

[135]

Phototaxis by amoebae of Dictyostelium discoideum

Amoebae of the cellular slime mould *Dictyostelium discoideum* move towards or away from unilateral light, depending on the light intensity (Häder & Poff 1979*a*, *b*, *c*; Hong, *et al.* 1981). Because of the small size of the organism and the relatively long wavelength to which it is sensitive, this organism presents a particular challenge for understanding the mechanism whereby light direction is detected.

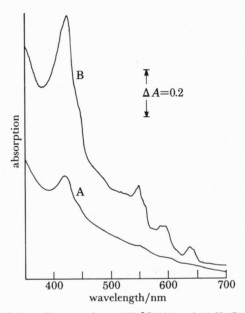

FIGURE 4. Absorption spectra of *Dictyostelium* amoebae at 23 °C (A) and 77 K (B). The sample consisted of 4×10^7 cells in 0.4 ml suspension buffer. The pathlength was approximately 0.3 cm. The spectra were measured by using a single-beam spectrophotometer on line with a small computer. Note the distinct absorption maximum at 640 nm in cells at 77 K, and the relatively low absorbance of cells at 23 °C.

Action spectra for both positive and negative amoebal phototaxis show a major peak at 405 nm, with secondary maxima at 440–520, 580 and 640 nm, Of these, the action maximum at 640 nm is of particular interest. This action peak has been associated with an absorption maximum *in vivo* at 640 nm, which may be easily seen in cells at 77 K (figure 4*a*). It is not readily evident that any of the known mechanisms are sufficient to establish any substantial gradient of 640 nm light in *Dictyostelium* amoebae.

1. The ability of a lens to focus light decreases rapidly as the diameter of the lens approaches the wavelength of light. The diameter of an amoeba is approximately 10 μm but is highly irregular, whereas the diameter of the more regular pseudopodium is about 1 μm. Clearly, 640 nm (0.64 μm) light is perceived by the amoeba. However, at this wavelength the amoeba should be relatively ineffective as a lens.

2. Scattering is, in general, inversely proportional to some power of the wavelength. Scattering could be quite significant in establishing a light gradient in the blue, but would be much less effective at 640 nm.

3. Establishing any significant gradient of light intensity by absorbance screening is unlikely, given the very low absorbance at 640 nm for cells at 22 °C (figure 4), and the very short path length (10 μm) through an amoeba.

Both scattering and absorbance screening may in fact operate to produce a significant gradient in light intensity in *Dictyostelium* amoebae. It should be noted, however, that any lens effect would operate to diminish the gradient established by scattering and absorbance screening.

The fact that none of the mechanisms for producing a light gradient is substantial in *Dictyostelium* amoebae exposed to unilateral 640 nm light raises a major question concerning the extent of the gradient required. How large must the ΔI be between the proximal and distal sides for an organism to measure the difference? Surely a ΔI of 50 % would be sufficient, but would a ΔI of 10^{-6} % suffice? It should be possible to calculate, for a given number of excitations, the difference on the distal and proximal sides necessary for statistical significance. Note that such a calculation would be valid only for a particular number of excitations or for a particular fluence rate. Unfortunately, this calculation presupposes a knowledge of the 'noise' level for the particular reaction or pathway modulated by light.

In summary, none of the known mechanisms is obviously sufficient to establish a significant gradient of 640 nm light in *Dictyostelium* amoebae.

Phototropism in monocot seedlings

For many monocots, the dose–response curve for phototropism is very complex, typically showing at least three separable components, which have been termed first positive phototropism, first negative phototropism, and second positive phototropism (figure 5). Considerable evidence has been accumulated that separable mechanisms are involved in these responses (Zimmerman & Briggs 1963), although the nature of the mechanisms and their difference are largely unknown. These differences could be based on different photoreceptor pigments, different response mechanisms, or different mechanisms for detecting light direction in the first and second positive phototropic responses.

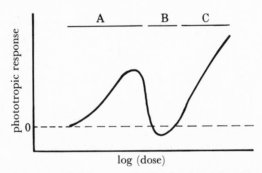

FIGURE 5. Idealized dose–response curve for phototropism by the shoot of a monocot seedling. A, B and C represent first positive phototropism, first negative phototropism and second positive phototropism, respectively.

In spite of the evidence that the first and second positive phototropic responses differ considerably, it has frequently been tempting to extrapolate from one to the other, equating the two responses. Such an approach may in fact delay an understanding of the phenomenon of phototropism. Perhaps the best example of this is the evaluation of the role of screening in phototropism.

It has clearly been established that screening is involved in second positive phototropism, i.e. that the high absorption in the primary leaf within the coleoptile shades the distal side of the coleoptile such that an intensity gradient is established between the proximal and distal sides. This has been demonstrated through a number of experiments: (1) if the primary leaf is removed, the screening is dramatically reduced and phototropism decreased; (2) if the primary leaf is replaced by an absorbing dye, the screen and phototropism are regenerated (Brauner 1955; Bunning *et al.* 1953); (3) if one treats seedlings with SAN 9789, an inhibitor of carotenoid biosynthesis, both the screen in the coleoptile and primary leaf and phototropic sensitivity are substantially reduced (figures 6 and 7). Thus one may conclude that light direction is detected in second positive phototropism through the mechanism of screening.

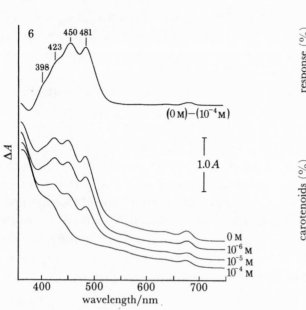

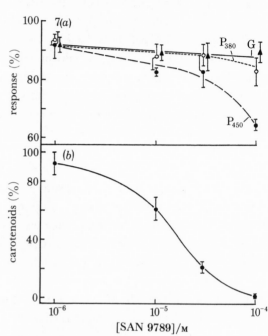

FIGURE 6. Absorption spectra of corn [maize] seedlings germinated with water or different concentrations SAN 9789. The sample consisted of 0.5 g of seedling tips including the primary leaf homogenized in 0.5 ml distilled water. Spectra were measured by using a single-beam spectrophotometer on line with a small computer. (Taken from Vierstra & Poff (1981).)

FIGURE 7. The effect of SAN 9789 on carotenoid accumulation, phototropism and geotropism in corn seedlings.
(*a*) The effect of SAN 9789 on phototropism and geotropism of corn seedlings. Geotropic bending (G; ▲) relative to control seedlings germinated in distilled water was measured after 3 h geotropic stimulus. Phototropic bending to 380 nm (○) and 450 nm (●) light, relative to control seedlings germinated in distilled water, was measured after a 3 h phototropic stimulus. The vertical bars represent ±1 standard error. Each point represents from four to six independent experiments comparing ten seedlings treated with SAN 9789 with ten control seedlings.
(*b*) The effect of SAN 9789 on carotenoid accumulation in corn seedlings. The carotenoid content of SAN 9789 treated seedlings was determined from the absorbance at 481 nm of 0.5 g homogenized seedling tips and compared with that of control seedlings. The vertical bars represent ± one standard deviation.

It is not equally clear that screening is involved in first positive phototropism. Moreover, it should be noted that a relatively high-absorbance screen is available in the coleoptile only in the primary leaf (figure 8). In contrast, the absorbance of the tip of the coleoptile above the primary leaf is low and may not be consistent with screening as a mechanism for the detection of light direction but may be consistent with a refraction mechanism.

Considerable attention has been given to the tip and base responses of *Avena* and to the correlation of the 'tip response' with first positive phototropism (Thimann & Curry 1961;

Curry 1969). Dennison (1979), however, argues that the distinction between the 'tip response' and 'base response' is morphological and spurious. However, these arguments seem only to be directed toward the bending response and do not negate the observations that only the tip is sensitive at low light doses.

If the tip of the coleoptile is indeed responsible for the first positive phototropism, then the possible role of the refraction mechanism for the detection of light direction should be carefully examined given a low transverse absorbance in the coleoptile tip. If refraction is the mechanism for the detection of light direction in the tip, and if the photoproducts are not easily diffusible,

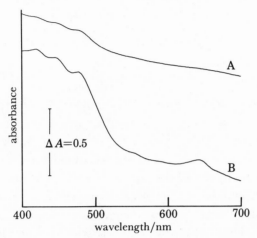

FIGURE 8. Absorption spectra measured across the shoot of a corn seedling. Spectrum A was measured through the tip of the coleoptile above the primary leaf. Spectrum B was measured 0.5 cm below the tip of the coleoptile, where the primary leaf is present. The seedlings were grown for 5 days in darkness with 1 h red light each day. Spectra were measured by using a single-beam spectrophotometer on line with a small computer.

then one would expect a 'negative' response after the first positive response. This would occur after the photoreceptor-response mechanism is saturated on the distal side and before saturation on the proximal side. Thus the extent and perhaps the existence itself of the 'negative' response should be fluence-dependent and would be expected to be more extreme at higher fluence rates where saturation would be more severe. That such a fluence-dependent first negative phototropism is indeed observed may be purely chance and should not be accepted as evidence that refraction is involved in first positive phototropism. However, this should be sufficient to stimulate a closer examination of the mechanism whereby light direction is measured in first positive phototropism.

This work was supported by the U.S. Department of Energy under contract no. AC02-76ERO-1338.

References

Brauner, L. 1955 Über die Funktion der Spitzenzone beim Phototropismus der *Avena*-koleoptiles. *Z. bot.* **43**, 467–498.

Bunning, E., Dorn, J., Schneiderhohn, G. & Thorning, J. 1953 Zur Funktion von Lactoflavin und Carotin beim Phototropismus and bei Licht bedingten Wachstums beeinflussungen. *Ber. dt. bot. Ges.* **66**, 333–340.

Curry, G. M. 1969 Phototropism. In *The physiology of plant growth and development* (ed. M. B. Wilkins), pp. 241–273. New York: McGraw-Hill.

Dennison, D. 1979 Phototropism. In *Encyclopedia of plant physiology* (ed. W. Haupt & M. Feinleib), vol. 7 (*Physiology of movements*), pp. 506–566. Berlin, Heidelberg and New York: Springer-Verlag.

Feinlieb, M. 1980 Photomotile responses in flagellates. In *Photoreception and sensory transduction in aneural organisms* (ed. F. Lenci & G. Colombetti), pp. 45–68. New York and London: Plenum.

Häder, D.-P. & Poff, K. 1979*a* Light-induced accumulations of *Dictyostelium discoideum* amoebae. *Photochem. Photobiol.* **29**, 1157–1162.

Häder, D.-P. & Poff, K. 1979*b* Photodispersal from light traps by amoebae of *Dictyostelium discoideum*. *Expl Mycol.* **3**, 121–131.

Häder, D.-P. & Poff, K. 1979*c* Inhibition of aggregation by light in the cellular slime mold *Dictyostelium discoideum*. *Arch. Microbiol.* **123**, 281–285.

Hong, C., Häder, M., Häder, D.-P. & Poff, K. 1981 Phototaxis in *Dictyostelium discoideum* amoebae. *Photochem. Photobiol.* **33**, 373–377.

Thimann, K. & Curry, G. 1961 Phototropism. In *Light and life* (ed. W. McElroy & B. Glass), pp. 646–670. Baltimore, Maryland: The Johns Hopkins Press.

Vierstra, R. & Poff, K. 1981 Role of carotenoids in the phototropic response of corn seedlings. *Pl. Physiol.* **68**, 798–801.

Zimmerman, B. & Briggs, W. 1963 A kinetic model for phototropic responses of oat coleoptiles. *Pl. Physiol.* **38**, 253–261.

Discussion

S. OBRENOVIĆ (*Institute for Biological Research 'Siniša Stanković', Belgrade, Yugoslavia*). Concerning the effect of norfluorazone on carotenoids as screening pigments and on the phototropic response, I wish to ask whether Dr Poff has measured its effect on the presumed photoreceptor for blue light, the flavin–cytochrome complex. We have measured the light-induced absorbance change in the 50 000 *g* pellet fraction from corn coleoptiles and found that norfluorazone does affect it. The effect of norfluorazone on the phototropic reaction was established by Dr Konjevic only in light-grown plants (*Phaseolus aureus* Roxb.) and not in etiolated ones. A comparable effect was obtained on the light-induced absorbance change *in vivo*. So it seems that norfluorazone can affect directly the presumed photoreceptor for blue light, and thus its effect on the phototropic reaction can not be unequivocally ascribed to the lack of carotenoids.

K. L. POFF. No, we have not measured the effect of norfluorazone on the blue-light-induced absorbance changes. The results just described are interesting and may suggest an effect, whether direct or indirect, on the photoreceptor pigment itself. In our experience, the specificity of inhibitors is dependent upon the concentration used so one should be cautious in extrapolating results from one experiment to another.

Several lines of evidence support the conclusion that the carotenoids function primarily as a screening pigment. (1) Fluence response curves with and without norfluorazone extrapolate to zero response at the same fluence. This is probably not compatible with an effect of the inhibitor directly on the photoreceptor pigment. (2) Experiments where the primary leaf is removed from the coleoptile result in a decreased phototropism. The addition of a dye in place of the primary leaf regenerates the phototropic response (Bunning 1953; Brauner 1955).

I agree that one should be cautious in the use of inhibitors remembering that few if any are specific at all concentrations. However, in this case, I believe that the data support the conclusion that the carotenoids function as a screening pigment in corn coleoptiles and that norfluorazone inhibits phototropism through the decrease of that screen.

S. OBRENOVIĆ. What are Dr Poff's views on the involvement of phytochrome in the phototropic reaction?

K. L. POFF. Although it is clear that phytochrome is related in some way to the phototropic response, perhaps potentiating the response, I would be extremely hesitant to propose a specific role of phytochrome. No attempt has been made in this paper to include phytochrome because

I am aware of no evidence suggesting that phytochrome directly establishes the light gradient in etiolated corn shoots. However, we must understand the role of phytochrome in phototropism before we truly understand the mechanism of phototropism.

W. HAUPT (*Institut für Botanik und Pharmazeutische Biologie, Erlangen, F.R.G.*). I have a comment on the expected composite action spectrum, containing characteristics of photoreceptor and screening pigments. In the temporal gradient, due to periodic screening (e.g. in *Euglena*), I would not expect any major contribution of the photoreceptor's absorption spectrum to the action spectrum. The photoreceptor is adapted to the steady-state intensity and the response depends on the proportional step-down signal, irrespective of the steady-state level. The size of this step-down signal is solely a function of the absorption in the screening pigment (given the fact that the photoreceptor can absorb at all). This comment does not concern the situation with spatial gradients.

K. L. POFF. That is a very interesting suggestion. It would appear that in such a system, given absorption by the photoreceptor pigment, adaptation to the 'unscreened' light, and a response proportional to the step-down signal, the contribution by the photoreceptor pigment to the action spectrum would be minimal. Thus the major contribution of the photoreceptor pigment to the action spectrum would be to set the wavelength limits for the response.

R. D. FIRN (*Department of Biology, University of York, U.K.*). I wonder whether Dr Poff might not be underestimating the contribution of diffusion and light scattering to the creation of a light gradient across a coleoptile. Some studies we have recently made on light gradients in hypocotyls suggests that these factors are important and it is evident from Dr Poff's data that coleoptiles lacking the carotenoid screening pigments still show 60 % of the normal phototropic response.

K. L. POFF. As I indicated, screening may be accomplished through scattering or absorption either by the photoreceptor pigment itself or by a second pigment. Of these, the least complicated to manipulate experimentally is absorptive screening by a second pigment. One must not forget that scattering and absorption must both be present inherently. A quantitative study of the relative importance of each of these factors has not yet been made in any system but would be a significant contribution.

Phil. Trans. R. Soc. Lond. B **303**, 489–501 (1983)

Printed in Great Britain

Photoperception and de-etiolation

By H. Mohr and E. Schäfer

Biological Institute II, University of Freiburg, Schänzlestrasse 1, D-78 Freiburg, F.R.G.

In seedlings or sprouts of higher plants, photomorphogenesis is the strategy of development if and as long as abundant light is available, and scotomorphogenesis (etiolation) is the developmental strategy of choice as long as light is not yet, or no longer, available. The transition from scotomorphogenesis to photomorphogenesis (called de-etiolation) can be considered a process in which a single, well defined environmental factor causes a plant to change its pattern of gene expression. The present article focuses on the question: what is the photosensory system, including photoreception and signal transduction, through which a plant can detect those light conditions that justify the (gradual) shift from scotomorphogenesis to photomorphogenesis, i.e. de-etiolation, which implies a strong and partly irreversible investment of matter and energy? The significance of phytochrome for signal reception, the mode of signal expression, and the time course of signal transduction in phytochrome-mediated responses are reviewed briefly. The emphasis is on amplification of the phytochrome signal by red, blue and ultraviolet light (measured as responsivity amplification) because these recent findings may lead to a better understanding of the responses of plants under natural light conditions.

Introduction

Every seedling or sprout of terrestrial plants is genetically endowed with the ability to follow two different strategies of development, depending on the ambient light conditions. These are photomorphogenesis and scotomorphogenesis (figure 1). Photomorphogenesis is the strategy of development if and as long as light is available, and scotomorphogenesis (etiolation) is the developmental strategy of choice as long as light is not yet, or no longer, available. The adaptive value of having different strategies is obvious. Scotomorphogenesis is the appropriate strategy of survival under conditions where light is lacking, whereas photomorphogenesis is the appropriate strategy of development under conditions of light affluence.

The plants in figure 1 illustrate extreme cases (dark compared with full sunlight). The appearance of a plant under conditions where light is *sometimes* limiting may show features of etiolation (e.g. elongated internodes) even though photomorphogenesis dominates the scene.

The transition from scotomorphogenesis to photomorphogenesis (called de-etiolation) can be considered a process in which a single, well defined environmental factor causes a plant to change its pattern of gene expression (Mohr 1972).

Photomorphogenesis and scotomorphogenesis can be distinguished not only on the level of the organism but also in intracellular morphogenesis of organelles. As an example, in the presence of light a proplastid develops into a green mature chloroplast whereas development in darkness follows a different strategy, leading to an etioplast. Upon irradiation with white light, rapid 'de-etiolation' of the etioplast takes place, i.e. the light-mediated transformation of the etioplast into a mature chloroplast occurs.

De-etiolation is, in part at least, reversible (re-etiolation). An example of re-etiolation is

[143]

dark-mediated plastid 'senescence', which takes place if a (mustard) plant is kept in darkness for more than 2 days. Dark-mediated plastid senescence, including loss of pigment and breakdown of fine structure, can be prevented by red light pulses, which operate through phytochrome (Biswal *et al.* 1983). Another example of re-etiolation is the phytochrome-controlled synthesis of phytochrome. During phytochrome-mediated de-etiolation the synthesis of phytochrome becomes suppressed (Gottmann & Schäfer 1983). However, this suppression persists only as long as P_{fr} is present (V. Otto, personal communication).

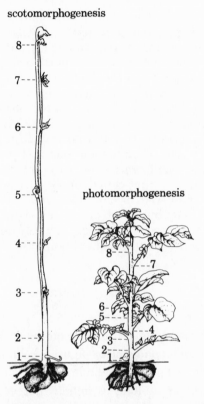

FIGURE 1. Illustrations of alternative developmental strategies. Genetically identical potato plants (*Solanum tuberosum* L.) were grown in the dark (scotomorphogenesis) or in natural daylight (photomorphogenesis) (after Pfeffer 1904). The numbers indicate the position of the corresponding leaves along the main axis to document the constancy of the phyllotactic pattern in light and dark (see Mohr 1978).

In order to respond properly in the delicate interplay between scotomorphogenesis and photomorphogenesis a plant has to sense the light conditions in its environment precisely. This paper will focus on the question: What is the photosensory system, including photoreception and signal transduction, through which a plant can detect those light conditions that justify the (gradual) shift from scotomorphogenesis to photomorphogenesis, i.e. de-etiolation, which implies a strong and partly irreversible investment of matter and energy?

SENSOR PIGMENTS

With regard to de-etiolation (excluding phototropism) higher plants possess three sensor pigments: phytochrome, a blue u.v.-A light photoreceptor ('cryptochrome') and a specific u.v.-B photoreceptor (see Mohr 1983). The action spectrum related to the latter photoreceptor shows a single intense peak at 290 nm and no action at wavelengths longer than 350 nm (Yatsuhashi *et al.* 1982). It could be that the u.v.-B photoreceptor occurs widely but has been overlooked so far owing to experimental difficulties in working with u.v.-B. In any case, at least some plants (e.g. milo, *Sorghum vulgare* Pers.) are obviously capable of sensing the light conditions throughout the Sun's spectrum, i.e. as far as the sunlight leads to electronic excitations (290–800 nm) (Drumm-Herrel & Mohr 1981).

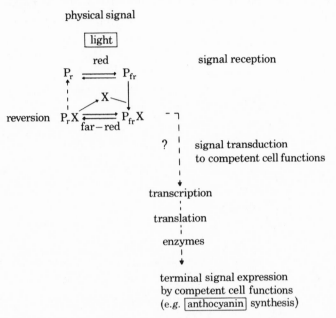

FIGURE 2. A scheme to illustrate the sequence of events occurring in the light-mediated synthesis of flavonoid compounds. The basic observation is that the environmental (physical) signal light leads to synthesis of particular pigment molecules, e.g. anthocyanin, in particular cells. The core of the explanatory argument is that terminal signal expression (i.e. the appearance of the final response, in the present case cyanidin, or flavone glucosides) is due to the induction of competent genes, i.e. to the onset of transcription. The premises that signal reception is taking place through phytochrome and that competent cells in a de-etiolating plant respond to the amount of P_{fr} (rather than to the P_{fr}/P_{tot} or P_{fr}/P_r ratio) are well substantiated (see Schmidt & Mohr 1982). The P_{fr} 'receptor', X, is largely hypothetical, and the mechanism of signal transduction from P_{fr} to the competent genes is still an enigma. On the other hand, the sequence of events between the induced appearance of RNA and the appearance of flavonoid compounds is well documented.

MODE OF SIGNAL EXPRESSION IN PHYTOCHROME-MEDIATED RESPONSES

As far as we know today, phytochrome is involved as a photoreceptor in all processes of de-etiolation in higher plants. For this brief review of what we know about the mode of action of phytochrome in bringing about de-etiolation we consider two 'biochemical model systems' of de-etiolation that have been investigated intensively in Freiburg, namely light-mediated synthesis of flavone glucosides in cell suspension cultures of parsely and light-mediated synthesis of juvenile anthocyanin in the epidermal cells of mustard (*Sinapis alba* L.) cotyledons. Regarding the 'molecular' mechanism of phytochrome action there is hardly any doubt that signal expression, i.e. the appearance of the flavonoid pigments, is due to the induction of competent genes (figure 2). Experimental evidence in favour of this scheme has been summarized repeatedly (see, for example, Hahlbrock *et al.* 1976; Schröder *et al.* 1979; Kreuzaler *et al.* 1983; Mohr 1982; Mohr & Schopfer 1977) and need not be repeated here. However, some recent data obtained in crucial experiments to test the above scheme (figure 2) further will be briefly presented.

1. The claim that phytochrome-mediated induction of enzymes of the flavonoid pathway is due to synthesis *de novo* of enzyme protein was confirmed by immunotitration studies on phenylalanine ammonia-lyase and chalcone synthase (figure 3). The data show that under all experimental conditions enzyme activity of an extract is proportional to the amount of immunoresponsive material. No indication was found with the present assay of enzymically inactive ᴸ ᐧ immunoresponsive material in extracts prepared from dark-grown seedlings.

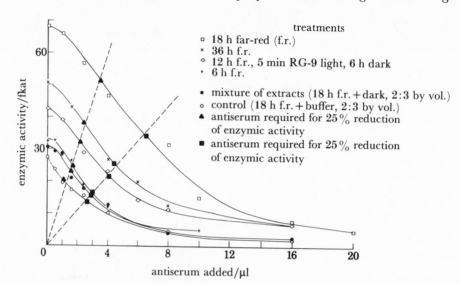

FIGURE 3. Immunotitration of chalcone synthase in extracts from mustard (*Sinapis alba* L.) cotyledons. Antiserum (kindly supplied by Professor K. Hahlbrock) was raised in rabbits by injections of highly purified chalcone synthase isolated from parsley cell suspension cultures (for details see Schröder *et al.* 1979). Control serum had no inactivating effect. Chalcone synthase activity is hardly detectable in extracts from dark-grown cotyledons. However, a strong phytochrome-mediated activity increase is observed in continuous far-red light with a peak around 18 h after the onset of light (36 h dark germination). The column headed 'treatments' designates the treatment of the seedling before extraction, and the preparation of mixed extracts to test for immunoresponsive material in an extract from dark-grown seedlings. The assay for chalcone synthase (see Heller & Hahlbrock 1980) was performed after the addition of antiserum, incubation (30 min at 30 °C, 20 min in the cold), and centrifugation (10 min at 12000 rev. min⁻¹). The broken lines show that the amount of antiserum required to reduce enzyme activity by 25 or 50% is directly proportional to the activity of the original extract. Results shown are per cotyledon pair. (Data obtained by R. Brödenfeldt.)

2. The Hahlbrock group (Kreuzaler *et al.* 1983) has shown in parsley cell suspension cultures that light-mediated changes in the amount of chalcone synthase mRNA coincides with the light-mediated changes of chalcone synthase synthesis *in vivo* and *in vitro*. The data are consistent with the hypothesis that induction of chalcone synthase by light is due to a transient increase in the rate of synthesis of chalcone synthase mRNA.

3. Regarding de-etiolation of the plastid, i.e. etioplast–chloroplast photoconversion, it was shown by Apel (1979) in barley and by Tobin (1981) in duckweed that light (via phytochrome) mediates specifically the appearance in the cytosol of a prominent mRNA species which codes for the apoprotein of the light-harvesting chlorophyll *a/b* protein, LHCP, located within the thylakoid membrane. While continuous chlorophyll synthesis is required for the incorporation of the apoprotein into the thylakoid membrane, appearance of the mRNA in the cytosol is

exclusively phytochrome-controlled. It seems that in the absence of chlorophyll *a* the apoprotein of LHCP is unstable and subjected to rapid degradation (Apel & Kloppstech 1980).

Link (1982) has demonstrated with mustard seedlings, by means of hybridization of mRNA to cloned plastid DNA, that the phytochrome-mediated appearance of a prominent protein of the thylakoid membrane (32 kDa 'shielding' polypeptide of photosystem II) is due to a rise of the level of the pertinent mRNA transcribed from plastid DNA.

In brief: the mechanism of signal expression in 'positive' responses such as phytochrome-mediated flavonoid biosynthesis and plastid genesis seems to be settled: phytochrome causes – in some way or other – activation of competent genes.

Signal expression in 'negative' responses might become understandable along the same lines. As examples, rapid phytochrome-mediated *decreases* in mRNA activities have been observed in the phytochrome-mediated decline of synthesis of protochlorophyllide oxidoreductase (Apel 1981) and phytochrome apoprotein (Gottmann & Schäfer 1983). Presumably, phytochrome causes – in some way or other – the inactivation of competent genes.

SIGNAL PERCEPTION IN PHYTOCHROME-MEDIATED RESPONSES

In crucial experiments on phytochrome-mediated anthocyanin synthesis in mustard cotyledons, evidence was obtained that a mustard seedling responds to the amount of P_{fr} and not to the P_{fr}/P_{tot} ratio (Schmidt & Mohr 1982). Thus, the unsolved problem is signal transduction. At least in flavonoid synthesis and the synthesis of some plastidal proteins, 'signal transduction' means the transduction of the P_{fr} signal to the competent genes. Unexpected data were obtained in studies of the time course of signal transduction and with regard to signal amplification during transduction.

TIME COURSE OF SIGNAL TRANSDUCTION IN P_{fr}-MEDIATED RESPONSES

It was found in phytochrome action on chlorophyll (Chl) synthesis (Oelze-Karow & Mohr 1982) as well as in phytochrome-mediated anthocyanin synthesis (Schmidt & Mohr 1983) that P_{fr} operates in two steps. The principal findings – formulated for phytochrome action on chlorophyll synthesis – are the following: a brief red light pretreatment (pulse) operating through phytochrome stimulates the synthesis of Chl *a* and *b* in milo (*Sorghum vulgar* Pers.) shoots placed in continuous saturating white light. The red effect is fully reversible by a far-red (756 nm) light pulse for 45 min. Thereafter the escape from reversibility is fast, being completed within 2 h. The major finding has been that physiologically active phytochrome is required continuously during these first 45 min if the onset of the loss of photoreversibility is to begin 45 min after the red light treatment. Thus, the initial action of P_{fr} consists of two distinct processes: the first is to overcome the lag before signal transduction; the second is the actual signal transduction (probably to the pertinent genes). The duration of the lag before the onset of signal transduction depends on the level of P_{fr} established by the initial light pulse. The duration increases with increasing P_{fr} levels from non-detectable to 45 min. Above approximately 15% P_{fr} ($P_{fr}/P_{tot} \approx 0.15$), the duration of the lag before signal transduction remains constant at 45 min. We suggest explaining these novel findings in terms of interaction of large ligands (P_{fr}) with lattice-like chains (McGhee & von Hippel 1974), following in principle the argument that we have advanced previously to account for threshold phenomena in P_{fr} action (Mohr

& Oelze-Karow 1976) and for the shape of the P_{fr}-effect curve in light-mediated anthocyanin synthesis (Drumm & Mohr 1974). Corresponding phenomena are observed in light-mediated decrease of phytochrome content in the milo shoot. The same lag (45 min) was measured before the onset of destruction. Moreover, it was found that P_{fr} is required continuously to overcome the lag before the onset of loss (M. Sauter, personal communication).

SIGNAL AMPLIFICATION

It has been noticed (Mohr *et al.* 1979) that the effectiveness of P_{fr} in mediating anthocyanin or phenylalanine ammonia-lyase synthesis in mustard cotyledons can be increased strongly by a light pretreatment of the seedling before competence. The effect (called 'increase of

TABLE 1. INDUCTION (OR LACK OF INDUCTION) OF ANTHOCYANIN SYNTHESIS IN THE MESOCOTYL OF MILO SEEDLINGS BY LIGHT OF DIFFERENT QUALITIES (AFTER DRUMM & MOHR 1978)

(Treatment was started 60 h after sowing; anthocyanin was measured 87 h after sowing.)

treatment	amount of anthocyanin (A_{510})
27 h dark	0
27 h white light†	1.85
27 h red light	0
27 h far-red light	0
3 h white light‡	0.19
3 h blue-u.v.	0.19
3 h white light + 5 min red light	0.19
3 h white light + 5 min 756 nm light	0.06
3 h white light + 5 min 756 nm light + 5 min red light	0.20
3 h blue-u.v. + 5 min red light	0.19
3 h blue-u.v. + 5 min 756 nm light	0.05
3 h blue-u.v. + 5 min 756 nm light + 5 min red light	0.19

Photoequilibria of the phytochrome are of the order of $\phi_{red} = 0.8$ and $\phi_{756} < 0.01$ (see Schäfer *et al.* 1975). A 5 min light pulse virtually suffices under the present circumstances to establish the photoequilibrium $\phi_\lambda = [P_{fr}]_\lambda/[P_{tot}]$.

† Xenon arc light, similar to sunlight, 250 W m^{-2}.

‡ Seedlings were kept in the dark for 24 h before extraction of anthocyanin.

effectiveness of P_{fr}' or 'increase of responsivity to P_{fr}', or in short, 'responsivity amplification', we prefer this last expression) can best be described in terms of 'signal amplification': a given amount of P_{fr} (signal) leads to a much higher response! The light that brings about 'signal amplification' in the mustard seedling is absorbed by phytochrome. Thus a light pretreatment, operating through phytochrome, leads to a strong albeit transient signal amplification in P_{fr}-mediated anthocyanin synthesis.

In mustard, phytochrome causes a signal amplification with regard to P_{fr}. In other cases, however, amplification of the P_{fr} signal is caused by light absortion in cryptochrome or in the u.v.-B photoreceptor, or both.

The first example is milo seedlings (*Sorghum vulgare* Pers., cv. Wieder-hybrid). The mesocotyl of the milo seedling does not produce anthocyanin in complete darkness. As described originally by Downs & Siegelman (1963) even long-term red or far-red light does not lead to any anthocyanin synthesis, whereas white light causes rapid pigmentation (table 1). In this case phytochrome can only act once a blue u.v. light effect has occurred. On the other hand, the expression of the blue u.v. light effect is controlled by phytochrome. In experiments with dichromatic irradiation, i.e. simultaneous irradiation with two kinds of light to modulate the level of P_{fr} strongly on a constant background of blue u.v. light, it was found that the blue u.v. light photoreaction as such is not affected by the presence or virtual absence of P_{fr} (Drumm

TABLE 2. INDUCTION (OR LACK OF INDUCTION) AND REVERSION OF ANTHOCYANIN SYNTHESIS IN THE COLEOPTILES OF *TRITICUM AESTIVUM* L. CV. SCHIROKKO, WITH LIGHT OF DIFFERENT QUALITIES (FOR DETAILS SEE DRUMM-HERREL & MOHR 1981) AND DIFFERENT LENGTHS OF TIME

(At the time indicated by a point (●) seedlings received either a saturating r.l. or RG-9 light pulse. Anthocyanin contents were measured at the age of 98 h. R.l., red light, 6.7 W m^{-2}, $\phi = 0.8$; RG-9 light, long-wavelength far-red light, 10 W m^{-2}, $\phi < 0.01$. As a gauge for responsivity to phytochrome we consider the extent of the reversible response, ΔR, defined by $\Delta R = R_1 - R_2$, where R_1 is the response obtained if the light treatment is terminated with a saturating red light pulse, and R_2 is the response obtained if the light treatment is terminated with a saturating long-wavelength far-red (RG-9) light pulse to return almost all P_{fr} to P_r (for definition of ϕ_λ see legend to table 1). (After Mohr & Drumm-Herrel (1983).)

	amount of anthocyanin (A_{546})		
treatment†	R_1	R_2	ΔR
5 d D (dark)		0.004	
2 d D + 10 h s.l.● + 40 h D	**0.169**	0.043	**0.126**
2 d D + 12 h w.l.● + 38 h D	0.092	0.030	0.062
3 d D + 2 h D● + 24 h D	0.004	0.004	0.000
3 d D + 2 h r.l.● + 24 h D	0.004	0.004	0.000
3 d D + 2 h b.l.● + 24 h D	0.004	0.004	0.000
3 d D + 2 h u.v. (WG345)● + 24 h D	0.008	0.004	0.004
3 d D + 2 h u.v. (PG218)● + 24 h D	0.012	0.004	0.008
2 d D + 12 h f.r.● + 38 h D	0.005	0.005	0.000
2 d D + 12 h r.l.● + 38 h D	0.004	0.004	0.000
2 d D + 12 h b.l.● + 38 h D	0.010	0.004	0.006
2 d D + 12 h u.v. (WG345)● + 38 h D	0.021	0.005	0.016
2 d D + 12 h u.v. (PG218)● + 38 h D	**0.160**	0.071	**0.089**
3 d D + 2 h u.v.-B● + 24 h D	0.033	0.012	0.021

† S.l., sunlight (changing fluxes due to clouds); w.l., fluorescent white light, 12000 lx; b.l., blue light, 7 W m^{-2}; u.v. (WG345), u.v.-A, 9.3 W m^{-2}; u.v. (PG218), u.v.-A with a small amount of u.v.-B, 12.6 W m^{-2}; u.v.-B, λ_{max} at 310 nm, 3 W m^{-2}.

& Mohr 1978). Our present interpretation of these facts is that phytochrome (P_{fr}) is always the effector molecule that causes anthocyanin synthesis through the activation of competent genes, whereas the blue–u.v. effect must be considered a transient 'responsivity amplification' (half-life approximately 6 h) (Drumm & Mohr 1978). In anthocyanin synthesis in the milo mesocotyl there is no P_{fr} signal transduction (and thus no responsivity) without the operation of a blue–u.v. photoreceptor.

The second example is wheat seedlings (*Triticum aestivum* L., cv. Schirokko). Anthocyanin synthesis in the wheat coleoptile takes place readily in natural sunlight and in fluorescent white light (table 2). Short-term and long-term treatments with red and far-red light proved ineffective. Surprisingly, blue and pure u.v.-A have very little effect, whereas u.v. light with a small content of u.v.-B is very effective in eliciting anthocyanin synthesis. After a u.v.

treatment the response was found to be controlled by phytochrome (figure 4). In fact, at least the major effect of the u.v. treatment is to establish responsivity to phytochrome. In contrast to milo, u.v. cannot be replaced by blue light in its responsivity amplification function. Thus the function of a separate u.v.-B photoreceptor is indicated.

The responsivity amplification for P_{fr} brought about by u.v.-B is only short-lived. It is completely lost if 12 h darkness is inserted between the u.v.-B treatment and a red light pulse (H. Drumm-Herrel, personal communication).

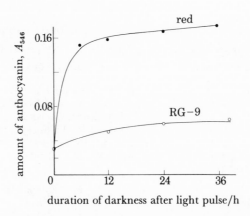

FIGURE 4. Anthocyanin accumulation in wheat (*Triticum aestivum* L.) coleoptiles in darkness after a u.v. pretreatment. The pretreatment (2 days dark, then 12 h u.v. (PG218) was terminated either with a saturating red light ($\phi_{red} = 0.8$, see table 1) or with a saturating RG-9 light pulse ($\phi_{756} < 0.01$, see table 1). (After Mohr & Drumm-Herrel (1983).)

The three species whose behaviours have been described briefly differ greatly with respect to the events underlying the same phenomenon, namely light-mediated synthesis of juvenile anthocyanin. Clearly, phytochrome is involved in the photoresponse in all cases. P_{fr}-mediated activation of competent genes seems to be at the core of the response in all cases. The different species vary with regard to the light-dependent 'mechanisms' that they use to establish responsivity towards P_{fr}. While the mustard seedling has some residual responsivity towards P_{fr} even in the dark and can perform a strong responsivity amplification via phytochrome, milo depends on light absorption in a blue u.v.-A photoreceptor (cryptochrome) to establish P_{fr} responsivity, whereas wheat even requires light absorption in a u.v.-B photoreceptor to become responsive to P_{fr} for anthocyanin synthesis.

A model for the process of responsivity amplification cannot be suggested at present. Increased responsivity was suggested to be due to a much more rapid signal transduction from P_{fr} to the responsive genes. This conjecture could not be supported in pertinent experiments (Schmidt & Mohr 1983). At present we are testing the hypothesis that responsivity amplification is due to an increase of some (proteinaceous?) factor in the cell nucleus that interacts with the P_{fr} signal.

This biochemical approach to the mechanism of signal transduction requires a response that takes place in many or most cells of an organ. We have chosen P_{fr}-mediated formation of Calvin cycle enzymes (ribulose 1,5-bisphosphate carboxylase, EC 4.1.1.39, and NADP-dependent glyceraldehyde 3-phosphate dehydrogenase, EC 1.2.1.13) in the shoot, predominantly in the primary leaf of the milo seedling. There is a small residual responsivity to P_{fr} that can be increased by blue–u.v. light approximately 60-fold within 6 h (figure 5). The high responsivity

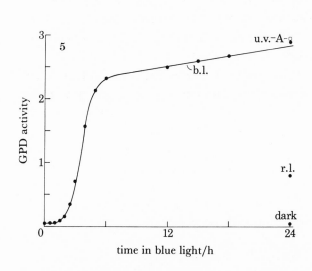

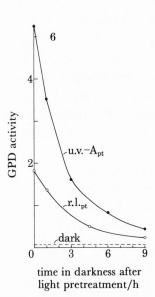

FIGURE 5. Blue-light-mediated responsivity increase in the milo (*Sorghum vulgare* Pers.) shoot with regard to phytochrome-mediated synthesis of NADP-dependent glyceraldehyde 3-phosphate dehydrogenase (GPD, EC 1.2.1.13). As a gauge for responsivity to phytochrome we consider the extent of the reversible response, ΔR (for definition see table 2). ΔR refers to the amount of total phytochrome, $[P_{tot}]$, present in the milo shoot at the time of the red and RG-9 light pulse treatment. We consider the ratio $\Delta R/[P_{tot}]$ the most precise gauge for the effectiveness of P_{fr} (or, expressed another way, for the responsivity towards P_{fr}). Dark-grown seedlings were pretreated with blue light (bl. 7 W m^{-2}) of different durations. At 72 h after sowing the pretreatment was terminated with a red or RG-9 light pulse. Until enzyme assay (96 h after sowing) seedlings were kept in darkness. U.v.-A (9.3 W m^{-2}) is as effective as bl., whereas red light (rl., 6.8 W m^{-2}) is much less effective. Dark, level of responsivity in dark-grown seedlings. (Data obtained by R. Oelmüller.)

FIGURE 6. Decrease of responsivity in darkness after a 24 h treatment with u.v.-A, terminated with an RG-9 light pulse to return almost all P_{fr} back to P_r (see table 2). The experimental system was the same as in figure 5. U.v.-A (9.3 W m^{-2}) and blue light (7 W m^{-2}) were adjusted so as to give approximately the same effect with regard to responsivity amplification (see figure 5). Red light (r.l., 6.8 W m^{-2}) of corresponding fluence rate is far less effective. Red or RG-9 light pulses were given after different durations of darkness but always at 72 h after sowing. Enzyme assay was performed 24 h after the light pulse, i.e. always 96 h after sowing. U.v.-A$_{pt}$, plants pretreated with 24 h of u.v.-A; r.l.$_{pt}$, plants pretreated with 24 h of red light. The low endogenous level of responsivity (dark) does not change during the period of examination. (Data obtained by R. Oelmüller.)

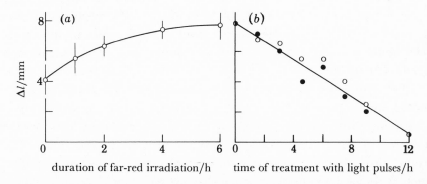

FIGURE 7. Experiments with mustard (*Sinapis alba* L.) seedlings. (*a*) Difference in hypocotyl length (Δl) caused within 24 h by a 5 min RG-9 light pulse compared with a 5 min red light pulse is plotted as a function of the duration of a continuous far-red light pre-treatment. The light pulses were always given 48 h after sowing. (*b*) A plot of Δl as a function of the dark interval between the pretreatment and the test light pulses. The pretreatment was 48 h dark plus 6 h white light. The white light was terminated with 5 min RG-9 light (o) or 5 min red light (●). Data obtained by B. Heim & C. Ebert.)

is only maintained as long as the blue or u.v.-A light is on. If we turn off the light, responsivity decays with a short half life (figure 6). In view of these data the level of responsivity to P_{fr} in the light must be considered a quasi-steady state that results from responsivity amplification by light and a light-independent decay reaction that decreases responsivity. Similar phenomena (increase of responsivity in light, decrease of responsivity after transferring to darkness) were also observed in control by phytochrome of hypocotyl elongation growth in mustard (figure 7). In mustard a phytochrome-mediated 'high-irradiance reaction', characterized by dependence on time (figure 7) and fluence rate (Beggs *et al.* 1981), leads to the responsivity amplification as tested by light pulses (figure 7). A contribution of a specific (i.e. non-phytochrome) blue light effect on responsivity could not be demonstrated (Beggs *et al.* 1981).

As shown in figure 7*b*, the decay of responsivity observed in the dark is independent of the P_{fr} level established at the onset of darkness. This suggests that continuous light or a rapid sequence of light pulses is required to cause an increase of responsivity.

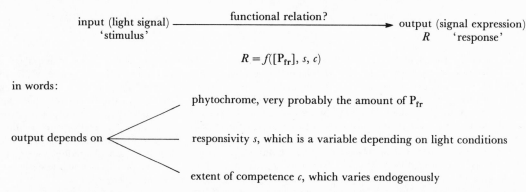

$$R = f([P_{fr}], s, c)$$

FIGURE 8. A scheme to account for the complexity of the functional relation between light signal and degree (extent) of response (R). Further explanation in text. The term 'competence' requires a comment: in developmental biology 'competence' means that a cell or a tissue is able to respond to a specific (inductive) stimulus with a specific response. As an example, an epidermal cell of a mustard cotyledon acquires competence towards P_{fr} with respect to anthocyanin synthesis approximately 27 h after sowing (25 °C). Competence disappears approximately 36 h later.

CONCLUSION

We conclude that de-etiolation, i.e. the gradual replacement of scotomorphogenesis by photomorphogenesis (see figure 1) is caused by phytochrome. In mustard the plant simply responds to the amount of P_{fr} established at the beginning of a dark period. However, the relation between amount of P_{fr} and the extent of response is a complex function because the responsivity of a plant towards P_{fr} depends on the light–dark conditions. Responsivity amplification by light can occur through phytochrome-mediated high-irradiance reaction and by light absorbed in a blue-u.v.-A light photoreceptor (cryptochrome) or in a u.v.-B photoreceptor, or both. Moreover, competence must also be considered. The changing spatial and temporal pattern of competence (Mohr 1978) is a crucial variable in any P_{fr} response function. Thus the output (extent of response) not only depends on the amount of P_{fr} but also on variables such as responsivity and competence (figure 8).

As an example, the perception of shade could be due to a reduced quantity of blue light below a leaf canopy – leading to a decrease of responsivity – *and* to a decrease of the amount of P_{fr} due to the strong depletion of red light, but relatively weak depletion of far-red light, by green leaves. Regarding light–dark transitions in adult plants we suggest that such transitions may not be sensed predominantly by a lowering of P_{fr} in darkness – because P_{fr} in totally de-etiolated plants seems to be quite stable (Gottmann & Schäfer 1983) – but by a relatively rapid decrease of responsivity.

In continuous light of constant quality and quantity an almost constant level of P_{fr} and a quasi-steady-state level of responsivity can probably be assumed after a couple of hours (see figures 5–7). However, under changing light conditions, in experiments with repeated light pulse treatments, or in light–dark experiments, the functional relation between the amount of P_{fr} and the extent of response becomes exceedingly complex. Thus quantitative experiments on 'molecular mechanisms' of signal transduction or signal expression should be performed either in continuous light or in darkness after single inductive light pulses (provided that responsivity is high enough).

An interpretation of action spectra obtained for de-etiolation responses under continuous irradiation will decisively depend on our knowledge about kinetics and wavelength dependence of responsivity towards P_{fr}. We suggest that the phenomena of 'high-irradiance reactions' are mainly due to responsivity amplification.

The research was supported by the Deutsche Forschungsgemeinschaft (SFB 206). We are greatly indebted to our research associates for placing unpublished data at our disposal.

References

Apel, K. 1979 Phytochrome-induced appearance of mRNA activity for the apoprotein of the light-harvesting chlorophyll a/b protein of barley (*Hordeum vulgare*). *Eur. J. Biochem.* **97**, 183–188.

Apel, K. 1981 The protochlorophyllide holochrome of barley (*Hordeum vulgare* L.). Phytochrome-induced decrease of translatable mRNA coding for the NADPH-protochlorophyllide oxidoreductase. *Eur. J. Biochem.* **120**, 89–93.

Apel, K. & Kloppstech, K. 1980 The effect of light on the biosynthesis of the light harvesting chlorophyll a/b protein. Evidence for the requirement of chlorophyll a for the stabilization of the apoprotein. *Planta* **150**, 426–430.

Beggs, C. J., Geile, W., Holmes, M. G., Jabben, M., Jose, A. M. & Schäfer, E. 1981 High irradiance response promotion of a subsequent light induction response in *Sinapis alba* L. *Planta* **151**, 135–140.

Biswal, U. C., Bergfeld, R. & Kasemir, H. 1983 Phytochrome-mediated delay of plastid senescence in mustard cotyledons: changes in pigment contents and ultrastructure. *Planta* **157**, 85–90.

Brüning, K., Drumm, H. & Mohr, H. 1975 On the role of phytochrome in controlling enzyme levels in plastids. *Biochem. Physiol. Pfl.* **168**, 141–156.

Downs, R. J. & Siegelman, H. W. 1963 Photocontrol of anthocyanin synthesis in milo seedlings. *Pl. Physiol.* **38**, 25–30.

Drumm, H. & Mohr, H. 1974 The dose response curve in phytochrome-mediated anthocyanin synthesis in the mustard seedling. *Photochem. Photobiol.* **20**, 151–157.

Drumm, H. & Mohr, H. 1978 The mode of interaction between blue (UV) light photoreceptor and phytochrome in anthocyanin formation of the *Sorghum* seedling. *Photochem. Photobiol.* **27**, 241–248.

Drumm-Herrel, H. & Mohr, H. 1981 A novel effect of UV-B in a higher plant (*Sorghum vulgare*). *Photochem. Photobiol.* **33**, 391–398.

Frosch, S., Bergfeld, R. & Mohr, H. 1975 Light control of plastogenesis and ribulose-bisphosphate carboxylase levels in mustard seedling cotyledons. *Planta* **133**, 53–56.

Gottmann, K. & Schäfer, E. 1983 Analysis of phytochrome kinetics in light-grown *Avena sativa* L. seedlings. *Planta* **157**, 392–400.

Hahlbrock, K., Knobloch, K.-H., Kreuzaler, F., Potts, J. R. M. & Wellmann, E. 1976 Coordinated induction and subsequent activity changes of two groups of metabolically interrelated enzymes. *Eur. J. Biochem.* **61**, 199–206.

Heller, W. & Hahlbrock, K. 1980 Highly purified 'flavanone synthase' from parsley catalyzes the formation of narigenin chalcone. *Archs Biochem. Biophys.* **200**, 617–619.

Kreuzaler, F., Ragg, H., Fautz, E., Kuhn, D. N. & Hahlbrock, K. 1983 UV-induction of chalcone synthase mRNA in cell suspension cultures of *Petroselinum hortense*. *Proc. natn. Acad. Sci. U.S.A.* **80**, 2591–2593.

Link, G. 1982 Phytochrome control of plastid mRNA in mustard (*Sinapis alba* L.). *Planta* **154**, 81–86.

McGhee, J. D. & von Hippel, H. P. 1974 Theoretical aspects of DNA–protein interactions: cooperative and non-cooperative binding of large ligands to a one-dimensional homogenous lattice. *J. molec. Biol.* **86**, 469–489.

Mohr, H. 1972 *Lectures on photomorphogenesis.* New York and Heidelberg: Springer-Verlag.

Mohr, H. 1977 Phytochrome and chloroplast development. *Endeavour* (N.S.) **1**, 107–114.

Mohr, H. 1978 Pattern specification and realization in photomorphogenesis. *Bot. Mag., Tokyo* (special issue) **1**, 199–217.

Mohr, H. 1982 Phytochrome and gene expression. In *Trends in photobiology* (ed. C. Hélène, M. Charlier, T. Montenay-Garestier & G. Laustriat), pp. 515–530. New York: Plenum Press.

Mohr, H. 1983 Criteria for photoreceptor involvement. In *Techniques in photomorphogenesis* (ed. H. Smith & M. G. Holmes). London: Academic Press. (In the press.)

Mohr, H., Drumm, H., Schmidt, R. & Steinitz, B. 1979 The effect of light pretreatments on phytochrome-mediated induction of anthocyanin and of phenylalanine ammonia-lyase. *Planta* **146**, 369–376.

Mohr, H. & Drumm-Herrel, H. 1983 Coaction between phytochrome and blue/UV light in anthocyanin synthesis in seedlings. *Physiologia Pl.* (In the press.)

Mohr, H. & Oelze-Karow, H. 1976 Phytochrome action as a threshold phenomenon. In *Light and plant development* (ed. H. Smith), pp. 257–285. London: Butterworths.

Mohr, H. & Schopfer, P. 1977 The effect of light on RNA and protein synthesis in plants. In *Nucleic acids and protein synthesis in plants* (ed. L. Bogorad & J. H. Weil), pp. 239–260. New York: Plenum Press.

Oelze-Karow, H. & Mohr, H. 1982 Phytochrome action on chlorophyll synthesis – a study of the escape from photoreversibility. *Pl. Physiol.* **70**, 863–866.

Pfeffer, W. 1904 *Pflanzenphysiologie.* Leipzig: Engelmann-Verlag.

Schmidt, R. & Mohr, H. 1982 Evidence that a mustard seedling responds to the amount of P_{fr} and not to the P_{fr}/P_{tot} ratio. *Pl. Cell Envir.* **5**, 495–499.

Schmidt, R. & Mohr, H. 1983 Time course of signal transduction in phytochrome-mediated anthocyanin synthesis in mustard cotyledons. *Pl. Cell Envir.* **6**, 235–238.

Schröder, J., Kreuzaler, F., Schäfer, E. & Hahlbrock, K. 1979 Concomitant induction of phenylalanine ammonia-lyase and flavanone synthase mRNAs in irradiated plant cells. *J. biol. Chem.* **254**, 57–65.

Tobin, E. M. 1981 Phytochrome-mediated regulation of messenger RNAs for the small subunit of ribulose 1,5-bisphosphate carboxylase and the light-harvesting chlorophyll *a/b* protein in *Lemma gibba*. *Pl. molec. Biol.* **1**, 35–51.

Yatsuhashi, H., Hashimoto, T. & Shimizu, S. 1982 Ultraviolet action spectrum for anthocyanin formation in Broom Sorphym first internode. *Pl. Physiol.* **70**, 735–741.

Discussion

R. J. ELLIS (*Department of Biological Sciences, University of Warwick, U.K.*). I should like to make a comment concerning the way that we think of phytochrome action with respect to changes in gene expression, and especially to changes in transcription. There is now evidence for a number of species and from a number of laboratories that light acting via phytochrome causes a large increase in the rate of transcription of certain genes. The point that I wish to make is that the evidence that we have so far does *not* suggest that light acts to switch genes on. I say this because wherever it has been looked for it turns out that these genes are transcribed in plants raised from seeds in total darkness, albeit at lower rates. For example, it is difficult to detect the mRNA for the small subunit of ribulose bisphosphate carboxylase in dark-grown *Pisum* plants if 5 µg total RNA is analysed by dot–blot hybridization, but when 30 µg poly $(A)^+$ RNA is analysed by Northern blotting, this dark level of mRNA is readily seen; this was first shown by S. Smith in his Ph.D work (1980, University of Warwick), and subsequently Sasaki *et al.* (*J. biol. Chem.* **256**, 2315–2320 (1980)) showed that this dark mRNA is translatable *in vitro*. So I suggest that we must not think of light acting via phytochrome on transcription in an absolute sense; rather light has a *kinetic* role: it causes the amplification of processes already occurring in absolute darkness. I think that this is an important distinction to make, especially for those of us trying to devise ways of demonstrating a relevant effect of added phytochrome in isolated subcellular systems.

H. MOHR. I agree with Professor Ellis that evidence from thoses cases that have been analysed so far on the molecular level are compatible with the concept that phytochrome increases the rate of transcription rather than switches genes on. The conclusion that phytochrome causes the amplification – sometimes one hundred or more times – of molecular processes already occurring in absolute darkness is certainly compatible with the currently available evidence. However, this is not the only interpretation. An alternative explanation of the 'leakiness' in darkness would be that in a few cells of an organ or a cell suspension culture transcription of a particular gene takes place in complete darkness whereas in most cells phytochrome is required to switch the gene on. The existence of responsivity amplification as an essential feature of phytochrome action – as described in the present paper – is compatible with Professor Ellis's suggestion that light via phytochrome 'causes the amplification of processes already occurring in absolute darkness' albeit at a low rate. In pursuing this problem we have found that the time course of competence of different genes towards phytochrome must be considered. It was observed in plastid genesis as well as in flavonoid biosynthesis that a dark level of an enzyme becomes detectable only once the particular transcription (or the particular gene) has become competent to the action of phytochrome. Thus we believe that the appearance of 'dark leakiness' of a particular transcription process has to do with those molecular changes that lead to competence.

Phil. Trans. R. Soc. Lond. B **303**, 503–521 (1983)

Printed in Great Britain

Perception of shade

By M. G. Holmes

Radiation Biology Laboratory, Smithsonian Institution, 12441 Parklawn Drive,
Rockville, Maryland 20852, U.S.A.

Plants perceive shade by responding to both the fluence rate and to the spectral quality of the natural radiation environment. Changes in fluence rate are perceived by separate photoreceptors absorbing in both the blue and the red wavebands. The identity of the photoreceptor (or photoreceptors) responding to changes in the fluence rate of blue light is unknown (see Briggs, this volume). Physiological responses to changes in the fluence rate in the red waveband appear to be mediated through phytochrome. The relative roles played by the blue-light-absorbing photoreceptor and phytochrome in determining the response to changes in fluence rate varies between species and organs and is also dependent on the physiological age of the plant. Evidence is also presented that supports the concept that phytochrome functions to perceive the specific form of shade caused by surrounding competitive vegetation.

1. Introduction

To perceive shade, a plant must possess one or more photoreceptors that react to changes in the natural light environment. The reaction to a light stimulus provides a means by which information on the prevailing light régime can be transduced in a form which enables, or causes, appropriate physiological responses. With this approach it is inevitable that energy-transducing photoreceptors, such as chlorophyll, are excluded from the discussion and that the study centres on signal-transducing photoreceptors that may enable a plant to react appropriately to shade. The method used here is to compare the spectral properties of natural light with the known properties of signal-transducing photoreceptors.

2. Shading by vegetation

Plants growing within vegetation shade experience a greatly different radiation environment from those growing in open habitats. Green leaves absorb, reflect and transmit light selectively. Blue and red wavebands are absorbed strongly, green light is absorbed less strongly and far-red radiation is largely reflected or transmitted. These patterns of selective absorption and scattering have also been observed under vegetation canopies (see, for example, Federer & Tanner 1966; Vezina & Boulter 1966; Stoutjesdijk 1972 a, b; Holmes & Smith 1975; Holmes & McCartney 1976; Holmes 1981). The attenuating effects of a wheat (*Triticum aestivum* L.) canopy on natural daylight are shown in figure 1. Three spectral changes in daylight as it passes through the canopy are of known physiological significance. First, the quantity of photosynthetically active radiation (p.a.r.) (400–700 nm) is strongly reduced. This reduction can affect dry mass accumulation rate. Second, the reduced quantity of blue radiation may be of importance to the fluence-rate-dependent responses controlled by a blue-light-absorbing photoreceptor. The third spectral change that is known to be of physiological importance is the strong depletion of the red waveband and relatively weak depletion of the far-red waveband. This spectral change is detected by phytochrome.

3. PERCEPTION OF LIGHT QUALITY

To detect light quality, a plant must be able to compare the relative quantities of light in two or more wavebands (Smith 1981 *a*). This can be achieved by two methods. One possibility is the interaction of two separate photoreceptors whose mediated response depends on the relative amounts of light absorbed by each photoreceptor. There are several responses known to be controlled by two separate photoreceptors, but quantitative information on how the two photoreceptors interact is limited (Mohr 1980).

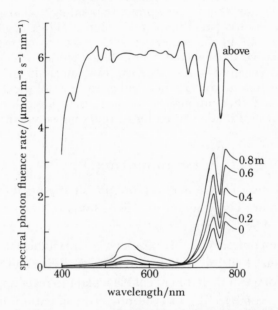

FIGURE 1. Spectral quality of radiation at various heights above ground level within a wheat (*Triticum aestivum* L.) canopy. The spectra were measured on 7 June 1973 with a crop height of 0.90–0.95 m under a clear sky (0/8 to 2/8 cloud cover, solar disc not obscured). (After Holmes & Smith (1977*b*).)

The alternative possibility for detecting light quality is for the plant to use a photochromic pigment, i.e. a single pigment existing in two forms that are interconvertible by light. Although photoreversible pigments which absorb in the blue–ultraviolet wavebands have been reported in fungi (Kumagai 1982; Löser & Schäfer 1980) and animals (Hamdorf *et al.* 1972), phytochrome is the only known photoreversible pigment in higher plants. There now exists extensive circumstantial evidence that the photochromic properties of phytochrome are used to detect the presence and extent of shading by other plants. The mechanism appears to depend on the absorption maxima of P_r and P_{fr} being centred around 660 and 730 nm respectively and the fact that the relative quantities of radiation in these wavebands varies under natural conditions.

(*a*) *Phytochrome action in light-grown plants*

The highest stem-elongation rates in dicotyledonous plants occur in the absence of light when either no P_{fr} has been formed (i.e. etiolation) or when the P_{fr} formed in light-grown tissue has been removed. Phytochrome action is therefore considered to be an inhibitory effect on stem extension because any light treatment that produces P_{fr} reduces the growth rate.

Most evidence indicates that P_{fr} is the active form of phytochrome in light-grown plants. This conclusion is based on arguments similar to those used by Hendricks *et al.* (1956) for 'bulk' and 'active' phytochrome (Jabben & Holmes 1983). Although the belief that P_{fr} is the active form of phytochrome has been questioned (Smith 1981 b; see also Smith, this volume), it will be necessary to find alternative explanations for the similarity between the action spectrum for inhibition of hypocotyl extension growth and the action spectrum for photoconversion of P_r to P_{fr} before alternatives can be considered.

Physiological experiments suggest that the P_{fr} formed in light-grown plants is very stable compared with the P_{fr} of dark-grown seedlings. An early indication of this was seen in the continued effectiveness of P_{fr} in the dark in inhibiting internode growth in *Phaseolus vulgaris* (Downs *et al.* 1957); removal of P_{fr} with far-red light during the dark period also removed the inhibition of growth that had persisted for up to 16 h after the end of the photoperiod. Similar indications of the continued effectiveness of the light received in the previous photoperiod have been monitored in the subsequent dark period by using linear displacement transducers (Lecharny & Jacques 1980; Morgan, Child & Smith, unpublished).

In *Sinapis alba*, the inhibitory effect of continuous red light on hypocotyl growth can be replaced by 5 min red light pulses interspersed by 55 min periods of darkness (Schäfer *et al.* 1981), indicating that there is no significant loss of P_{fr} over a 55 min period. However, it is clear that this P_{fr} is not completely stable because a 30 min irradiation (which gives complete photoconversion) followed by 23.5 h darkness is not as effective as 24 h continuous irradiation.

In contrast to the situation in dark-grown seedlings, phytochrome-mediated responses in light-grown dicotyledonous plants appear to be explainable on the basis of spectrophotometrically detectable phytochrome (Jabben & Holmes 1983). Phytochrome in light-grown plants is relatively stable and typically has a half-life of several hours (Butler *et al.* 1963; Koukkari & Hillman 1967; Clarkson & Hillman 1968; Wetherell 1969; Jabben & Deitzer 1978; Jabben 1980; Jabben *et al.* 1980; Atkinson *et al.* 1980; Heim *et al.* 1981; Kilsby & Johnson 1982; Rombach *et al.* 1982). Stable phytochrome derives from the apparently inactive labile pool that represents the 'bulk' phytochrome of dark-grown plants; the 'bulk' fraction is rapidly destroyed after photoconversion into P_{fr}. The phytochrome measured in plants growing in white light should not be considered to be stable unless it is demonstrated that rapid destruction does not occur after transferring the tissue to a light source that produces maximum conversion of P_r to P_{fr}. In other words, white light sources containing a high proportion of far-red light may gradually allow accumulation of labile phytochrome because the fraction held as P_r is protected from destruction.

(b) Phytochrome photoequilibrium

The variations in the relative proportions of red to far-red radiation that occur in Nature are known to cause changes in phytochrome photoequilibria *in vivo*. When etiolated plant tissue is placed on ice (to reduce the effects of phytochrome dark reactions) and exposed to a range of natural and artificial broadband spectra (Holmes & Smith 1977 a, b), a hyperbolic relation is observed between the red:far-red ratio of the incident radiation and the photoequilibrium established in the tissue (Smith & Holmes 1977) (figure 2). The most striking aspect of this relation is that phytochrome exhibits greatest sensitivity to the range of changes in red:far-red that are found in natural terrestrial environments. Terrestrial plants live under situations in which the red:far-red ratio varies between approximately 1.15 in full daylight and around 0.05 in dense vegetation shade (Holmes & Smith 1977*b*). It is precisely within this range that phytochrome is most sensitive to changes in light quality and would therefore act as an excellent perceiver of shade light quality.

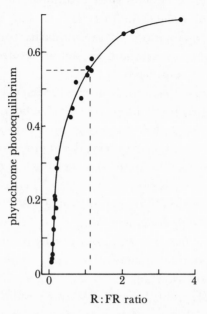

FIGURE 2. The relation between phytochrome photoequilibrium and red:far-red ratio of the incident poly-chromatic light. The broken line indicates photoequilibrium (*ca.* 0.54) established in natural daylight (red:far-red = *ca.* 1.15). Photoequilibria were measured in etiolated *Phaseolus vulgaris* hypocotyl hooks under both natural and artificial light sources. Both actinic irradiation and measurement of photoequilibria were done at 0 °C to reduce the effects of phytochrome dark (thermal) reactions. The figure demonstrates that phytochrome is most sensitive to changes in red:far-red ratio within the range found under natural terrestrial conditions. (After Smith & Holmes (1977).)

Two factors which might lead to deviations from the relation shown in figure 2 are the high fluence rates found in Nature and our lack of knowledge about the molecular environment of phytochrome. At high fluence rates, phytochrome exists for short periods as weakly absorbing intermediates between the P_r and P_{fr} forms (Kendrick & Spruit 1972), and under prolonged irradiation conditions the production of P_{fr} from P_r may become limited owing to the accumulation of the rate-limiting intermediate meta-Rb (Kendrick & Spruit 1976; Pratt *et al.* 1982; Spruit 1982). The molecular environment affects photoequilibrium (Mumford & Jenner 1971) and it is known that reaction rates *in vivo* are affected by both anaerobiosis (Kendrick & Spruit 1973) and by dehydration (Kendrick & Spruit 1974). However, in contrast to the situation in dark-grown tissue, most evidence points to the conclusion that phytochrome dark (i.e. non-photochemical) reactions play a relatively minor role in the establishment of photoequilibrium in light-grown plants at natural radiation levels (Jabben & Holmes 1982).

The average photoequilibrium established in green plants is lower than that in etiolated tissue because of the screening effect of chlorophyll (Holmes & Fukshansky 1979). Natural daylight, for example, establishes a photostationary state of about 0.54 in etiolated tissue whereas the average photostationary state in a green *Phaseolus* leaf illuminated on both surfaces is approximately 0.43. It should be noted, however, that although the photostationary state is lower in green tissue, the systematic relation between the red:far-red ratio of the incident light and phytochrome photoequilibrium is not affected (Morgan & Smith 1978a).

(c) Photoequilibrium and growth response

Red/far-red reversible control of stem elongation in light-grown plants was first demonstrated by Downs *et al.* (1957). They showed that 5 min far-red light given at the end of an 8 h photoperiod produced longer internodes in *Phaseolus vulgaris* than plants that had received 5 min far-red immediately followed by 5 min red light. These observations were important in that they not only demonstrated the effects of these spectral regions but also showed that the response is photoreversible. Similar responses have since been recorded in several other species (see, for example, Kasperbauer 1971; Lecharny & Jacques 1974; Vince-Prue 1973, 1977; Holmes & Smith 1977c).

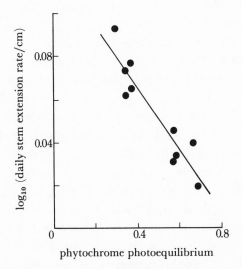

FIGURE 3. Relation between logarithmic stem extension rate in *Chenopodium album* and phytochrome photoequilibrium. Photoequilibria were derived from the curve in figure 2. The p.a.r. in all treatments was 100 μmol m⁻² s⁻¹. The figure demonstrates the systematic increase in elongation growth rate with decreasing P_{fr}/P_{tot} ratio. (After Morgan & Smith (1976).)

It became clear from observations (Holmes & Smith 1975), in which both the quantity and quality of light in the 400–700 nm waveband was held constant, that supplementary far-red light alone causes increased stem elongation rate. These experiments indicated that the decrease in red:far-red ratio below vegetation shade, which causes a decrease in the P_{fr}/P_{tot} ratio (Holmes & Smith 1975; Smith & Holmes 1977), causes a concomitant increase in elongation growth rate (Holmes & Smith 1975, 1977c). A more detailed analysis of this response was made by Morgan & Smith (1976), who used a range of photoequilibria resembling those existing in natural conditions. They observed an inverse linear relation between stem growth rate and the P_{fr}/P_{tot} ratio estimated to have been established during the daily photoperiod (figure 3). These results provide circumstantial evidence that phytochrome not only detects vegetation shade, but also measures the degree of shading.

The correlation between stem extention rate and estimated photoequilibrium depicted in figure 3 extends over the range $P_{fr}/P_{tot} = 0.71$–0.26 (Morgan & Smith 1978a). Technical restrictions made it impossible to produce lower photoequilibria at the background fluence rate of 100 μmol m^{-2} s^{-1} in the 400–700 nm waveband. Light-grown *Chenopodium rubrum* seedlings show a similar, but not identical, systematic relation between hypocotyl elongation rate and photoequilibrium over the range $P_{fr}/P_{tot} = 0.75$–0.17 (Ritter *et al.* 1981). Studies with light-grown *Sinapis alba* seedlings under monochromatic light show that this relation can hold down to P_{fr}/P_{tot} ratios as low as 0.03 (Holmes *et al.* 1982). Control experiments with chlorophyll-free plants indicated that photosynthesis was not involved in the response. A point of inflexion was observed at P_{fr}/P_{tot} lower than 0.03, where further reductions in P_{fr}/P_{tot} cause proportionally greater increases in hypocotyl elongation rate. It is noteworthy that no natural conditions are known where P_{fr}/P_{tot} ratios of less than 0.03 exist. It should also be pointed out that the growth responses of hypocotyls may not be the same as those of the growing internodes of mature plants.

The experiments described above were made at fluence rates that are known to establish photoequilibria within minutes under both the polychromatic (Holmes 1975; Smith & Holmes 1977) and monochromatic irradiation (Beggs *et al.* 1981; Jabben *et al.* 1982) conditions. At lower fluence rates, biphasic relationships between elongation response and predicted photo-equilibrium are observed. Both Satter & Wetherell (1968), studying *Sinningia*, and Vince-Prue (1973, 1977), studying *Fuchsia*, found maximal elongation rates under light sources that, on the basis of photochemical reactions alone, would have produced P_{fr}/P_{tot} values in the approximate range 0.30–0.40. A similar response was found in hypocotyls of light-grown *Chenopodium rubrum*, with the highest elongation growth rate at a photostationary state of approximately 0.35 (Holmes & Wagner 1981). However, when the experiments were repeated with a tenfold increase in fluence rate, the relation between elongation rate and photostationary state became approximately linear for the entire range ($P_{fr}/P_{tot} = 0.14$ to 0.69) studied. Holmes & Wagner concluded that the biphasic response was due to the relatively high effectiveness of phytochrome dark reactions at low fluence rates (Heim & Schäfer 1982). This explanation may also account for the biphasic responses observed with *Fuchsia* (Vince-Prue 1973, 1978) and *Sinningia* (Satter & Wetherell 1968), although species differences and slight effects of photosynthesis cannot be ruled out.

Light-grown plants appear to modulate their elongation rate in immediate response to changes in light quality. In *Chenopodium album*, stem extension rate is increased within 7 min of adding monochromatic far-red light to the background white fluorescent light (Morgan & Smith 1978 b). These rapid responses to radiation also occur after prolonged periods of darkness; *Vigna sinensis* L. responds within 30 min to monochromatic red light (Lecharny & Jacques 1980).

(d) Role of the light and dark periods

Although it is known that phytochrome controls elongation growth during the dark period (Downs *et al.* 1957) and during the light period (Holmes & Smith 1975), the relative effectiveness of these two periods is unknown. As outlined in the previous section, the qualitative responses to P_{fr} are the same in continuous light as in darkness after actinic irradiation. The responses differ in that some inhibitory effects of light are lost during long dark periods. In *Sinapis alba*, some inhibitory effect of light is lost during a 23.5 h dark period (table 1). If the dark period is reduced to 55 min, the inhibitory effect of 5 min red light pulses is as great as continuous red light. If the red light pulses are followed immediately by far-red light pulses, the response is approximately the same as with far-red light pulses alone, thereby indicating phytochrome involvement and a direct parallel with the characteristics of end-of-day responses.

Persistence of the effect of the previous light régime in darkness is also found in *Sinapis alba* plants 7 days old (table 1) and 2 weeks old (Child, Morgan & Smith, unpublished results). Similar observations have been reported by Lecharny & Jacques (1980).

In *Chenopodium rubrum*, a time-course study with both chlorophyll-free and normal green seedlings showed that a dark period of more than 12 h is required before a decrease in the inhibitory effectiveness of previous red light is observed (Holmes, Boden & Wagner, unpublished results). The response was red–far-red reversible, thereby implying phytochrome control.

TABLE 1. COMPARISON OF THE EFFECTIVENESS OF CONTINUOUS AND PULSED MONOCHROMATIC RED LIGHT ($\lambda_{max} = 654$ nm) ON INHIBITION OF HYPOCOTYL GROWTH IN LIGHT-GROWN *SINAPIS ALBA* L.

(Conditions: red light, 160 µmol m^{-2} s^{-1} for 5 min pulses, 3.3 µmol m^{-2} s^{-1} for 30 min pulse and 13 µmol m^{-2} s^{-1} for continuous light; far-red light (Schott RG9 + KG1) 25 µmol m^{-2} s^{-1} below 800 nm. (After Beggs *et al.* 1981; Schäfer *et al.* 1981).)

(*a*) *Seedlings* 54 h *old*

treatment	percentage inhibition
continuous red	66.8
24 × (5 min red + 55 min darkness)	66.9
24 × (5 min red + 5 min far-red + 50 min darkness)	5.9
24 × (5 min far-red + 55 min darkness)	10.2
30 min red + 23.5 h darkness	39.0

(*b*) *Plants* 7 *days old*

treatment	daily increase in hypocotyl length/mm
continuous red	1.4
24 × (5 min red + 55 min darkness)	1.1
24 × (5 min red + 5 min far-red + 50 min darkness)	3.6
24 × (5 min far-red + 55 min darkness)	3.5

Clearly, therefore, phytochrome can maintain elongation rate in darkness for several hours under otherwise constant conditions. In Nature, however, the relative roles of light and darkness appear to be strongly influenced by other factors. Using 16 h photoperiods, Morgan & Smith (1978*a*) found that 80% of the increased growth rate in *Chenopodium album* internodes caused by supplementary far-red light was produced by the daytime irradiation. It may be significant that the 20% attributable to the night period was produced over a shorter (8 h) period and at temperatures 5 °C lower than those used during the day.

(*e*) *Ecological variability*

The many changes in ontogeny elicited by simulated canopy light confer obvious adaptive advantages to species competing with others for light. For a ruderal plant such as *Chenopodium album*, the ability to redirect development toward vertical growth at the expense of lateral growth would enable the plant to intercept more light for photosynthesis. In many shaded habitats, such as within the understory vegetation of woods and forests, increased stem extension growth would be futile. There is evidence that species adapted to shaded environments respond differently from those that grow in open habitats.

Fitter & Ashmore (1974) showed that *Veronica persica* Poir., an annual weed of disturbed ground, radically altered its growth pattern in response to a simulated woodland light environment. In contrast to this response, *Veronica montana* L., a woodland perennial showed negligible change in its growth pattern under the same conditions.

In another study, Frankland & Letendre (1978) showed that the effects of supplementary far-red light on the growth pattern of the shade-tolerant plant *Circaea lutetiana* were not as marked as had been reported for plants adapted to open habitats. Morgan & Smith (1979) studied a wide range of species and concluded that the extent to which extension rate was modified by the prevailing red:far-red ratio was dependent on the natural habitat of the species (figure 4). Shade avoiders show marked increases in stem extension rate as the P_{fr}/P_{tot} level is decreased, whereas shade tolerators show little or no response to changes in photoequilibrium. This apparently systematic relation between natural species habitat and response to shade light quality would be of obvious value for growth and survival.

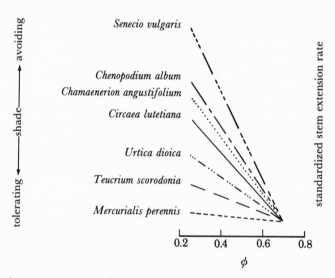

Figure 4. The relation between stem extension rate and phytochrome photoequilibrium. Photoequilibria were derived from the curve in figure 2. Stem extension data, expressed as logarithmic extension rate (i.e. $\log_{10}$ (growth per day)) were normalized to the value obtained at $P_{fr}/P_{tot} = 0.70$. The figure demonstrates the way in which the response to changes in red:far-red varies with the natural habitat of the species studied; the slopes of the lines for shade-avoiding species are very steep whereas those for the shade-tolerant species are very shallow. (After Morgan & Smith (1979) and Smith (1982).)

4. PERCEPTION OF LIGHT QUANTITY

Large gradients in the quantity of blue and red light exist within plant canopies. There are many examples in the literature of changes in growth and metabolism that are caused by variations in the quantity rather than the quality of light. Information is available on the effects of photon fluence rate variations at levels equivalent to sunlight (Warrington *et al.* 1978) down to the very low levels normally associated with vegetation shade (see, for example, Blackman & Wilson 1951*a*, *b*; Grime & Jeffrey 1965; Grime 1966).

Wavelength-sensitivity to changes in light quantity varies between species. In all species of light-grown plants studied in detail, it is changes in the quantity of either the blue, or the red, or both the blue and red wavebands that elicit the largest physiological responses. Two interesting examples with contrasting wavelength sensitivities are *Chenopodium rubrum* and *Sinapis alba*, both of which are successful competitors for light and both of which are very sensitive to changes in fluence rate. When studied at the stage of hypocotyl extension, *S. alba* is sensitive to the quantity of red light but shows negligible response to blue, whereas *C. rubrum* is highly sensitive to the quantity of blue light but is only slightly affected by changes in the quantity of red light. The different methods by which these two species perceive changes in the quantity of light will be analysed in the following sections.

(a) *The blue waveband*

The relative quantities of blue and red radiation in daylight are approximately the same at solar elevations higher than about 10° (see, for example, Holmes & Smith 1977*a*), and vegetation canopies attenuate blue and red light in approximately equal proportions (see, for example, Holmes & Smith 1977*b*). As red light is 50–100 times more effective than the blue waveband in photoconverting phytochrome (Butler *et al.* 1964; Pratt & Briggs 1966; Jabben *et al.* 1982), blue light will have only minor direct effects on phytochrome under natural conditions. However, there is substantial evidence for at least one specific blue-light-absorbing photoreceptor sensitive to changes in the quantity of light.

Various arguments have been put forward to support the concept that blue light responses can be explained on the basis of phytochrome action (Hartmann 1966; Schäfer 1975; Roth-Bejerano 1980). Most evidence, however, implies the existence of a separate blue-light-absorbing photoreceptor (b.a.p.) that operates in both light-grown and dark-grown plants.

Several of the experiments that have provided evidence of a specific b.a.p. have also indicated important factors that have to be considered when determining the contribution a b.a.p. may make towards the perception of shade. The first factor is that the relative effectiveness of blue and red light in controlling developmental growth varies between species (Vince 1956; Sale & Vince 1959). Meijer (1958, 1959) compared the effectiveness of equal fluence rates of monochromatic blue, red and far-red light on internode elongation in a variety of mature green plants. In two species, elongation rate was faster in blue than red; in two other species, elongation rate was slower in blue than red; in the fifth species studied, no significant difference was detected in elongation rate. He explained this difference between species on the basis of different fluence-rate dependences of two photoreceptors. Species differences are also common in light-grown seedlings. *Lactuca sativa*, *Cucumis sativa* and *Chenopodium rubrum*, for example, are relatively sensitive to blue light (Evans *et al.* 1965; Black & Shuttleworth 1974; Ritter *et al.* 1981) whereas light-grown *Sinapis alba* show a negligible response to blue light (Wildermann *et al.* 1978).

Internode elongation rate in *Vigna sinensis* is regulated by both blue and red light (Lecharny & Jacques 1980). Using a linear displacement transducer, they found a rapid response to both blue and red light. Compared with plants held in darkness, the cumulative effect of a 10 h irradiation treatment with blue light was promotive whereas the cumulative effect of red light was slightly inhibitory. However, when the blue and red light were given simultaneously, there was a marked inhibition of growth rate relative to the rate in darkness. Their results indicated a synergistic action of the blue and red wavebands. Synergistic effects of blue and red light have also been reported for malate formation in *Vicia faba* guard cells (Ogawa *et al.* 1978) and between blue and yellow light for control of stem growth in *Pisum sativum* (Elliott 1979).

It is clear from observations such as these that whereas studies with monochromatic light can increase our understanding of the individual photoreceptors that may be involved in the perception of shade, account must be made of possible interactions between photoreceptors under polychromatic irradiation conditions similar to those found in the natural environment.

As described above, sensitivity to light varies between species and organs and is also modified by age, stage and development, previous lighting conditions and, in some instances, the number of activated photoreceptors. In comparing the relative effectiveness of light quantity and quality in modifying development, it is necessary to reduce the number of variable factors for simplicity. To do this, the hypocotyl growth response to specific parameters of simulated shade light is compared in *Sinapis alba* L. and *Chenopodium rubrum* L.

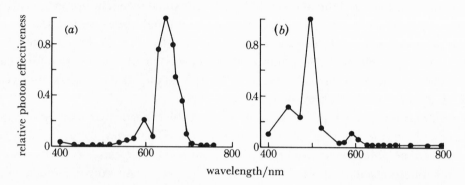

FIGURE 5. Action spectra for inhibition of hypocotyl elongation growth in (*a*) light-grown *Sinapis alba* seedlings and (*b*) light-grown *Chenopodium rubrum* seedlings. The spectra were divided from photon fluence rate curves (percentage inhibition against logarithmic photon fluence rate) for 50 % inhibition (*S. alba*) and 40 % inhibition (*C. rubrum*). The data are normalized to 1.0 at the wavelength of maximum inhibition. The action spectra illustrate differences between species in sensitivity to the blue and red wavebands. (Redrawn after Beggs *et al.* (1980) for *S. alba* and Holmes & Wagner (1982) for *C. rubrum*.)

S. alba, at the hypocotyl growth phase, responds to changes in both the quality and quantity of light in the red and far-red wavebands but shows negligible response to blue light (Figure 5), even when given as a supplement to red. In other words, the response in *Sinapis* allows a study to be made of the role played by phytochrome in perceiving shade without the additional complications produced by interacting effects of the b.a.p. *C. rubrum* is also sensitive to changes in light quality in red and far-red wavebands, but differs from *S. alba* in that it shows a relatively small response to the quantity of red or far-red light and a high sensitivity to the quantity of blue radiation (figure 5).

To determine the role played by the b.a.p. in the perception of shade it is necessary to reduce the number of variables in the actinic light. The known variables to which the plant may respond are phytochrome photoequilibrium (or concentration of P_{fr}), phytochrome cycling rate, the amount of p.a.r., the quantity of radiation in the red–far-red waveband and the quantity of radiation in the blue waveband.

The contributions of phytochrome photostationary state and cycling rate can be calculated, thereby making it possible to account for, or exclude, the contribution of these factors under either monochromatic or polychromatic radiation (Fukshansky *et al.* 1981). As a photochromic pigment, phytochrome can measure only two factors in the actinic radiations: these are the photons absorbed by P_r and the photons absorbed by P_{fr}. The product of these reactions depends on the photon fluence rate of the radiation, the quantum yield of the photoreactions and the extinction coefficients of P_r and P_{fr}. The photochemical reactions of phytochrome are described by the rate constants k_1 and k_2, i.e.

$$P_r \underset{k_2}{\overset{k_1}{\rightleftharpoons}} P_{fr},$$

where

$$k_1 = N_\lambda \, \epsilon_{r,\lambda} \, \phi_{r,\lambda}$$

and

$$k_2 = N_\lambda \, \epsilon_{fr,\lambda} \, \phi_{fr,\lambda}.$$

$\epsilon_{r,\lambda}$ and $\epsilon_{fr,\lambda}$ are the extinction coefficients of P_r and P_{fr} at wavelength λ; $\phi_{r,\lambda}$ and $\phi_{fr,\lambda}$ are the quantum yields of P_r and P_{fr} phototransformation at wavelength λ; N_λ is the photon fluence rate at wavelength λ.

Under polychromatic light, the rate constants k_1 and k_2 become

$$k_1 = \int_{\lambda_1}^{\lambda_2} N_{s,\lambda} \epsilon_{r,\lambda} \phi_{r,\lambda} \mathrm{d}\lambda$$

and

$$k_2 = \int_{\lambda_1}^{\lambda_2} N_{s,\lambda} \epsilon_{fr,\lambda} \phi_{fr,\lambda} \mathrm{d}\lambda,$$

where $N_{s\lambda}$ is the spectral photon fluence rate.

TABLE 2. SUMMARY OF THE ACTINIC PROPERTIES OF DAYLIGHT AND CANOPY LIGHT SPECTRA, AND THEIR CALCULATED EFFECTS ON PHYTOCHROME PHOTOTRANSFORMATION KINETICS (AFTER HOLMES ET AL. 1982)

parameter	daylight	canopy light
k_1/min^{-1}	132.1	1.36
k_2/min^{-1}	83.6	3.75
$(k_1+k_2)/\mathrm{min}^{-1}$	215.7	5.11
$k_1/(k_1+k_2)$	0.61	0.27
red:far-red ratio†	1.14	0.10
equivalent monochromatic wavelength/nm	687	701
equivalent monochromatic photon fluence rate/(μmol m^{-2} s^{-1})	452	17.7

† 10 nm bandwidth centred at 660 and 730 nm.

It should be noted that the photostationary state (i.e. $k_1/(k_1+k_2) = \phi$) is not the same as the P_{fr}/P_{tot} ratio established *in vivo* as a result of combined phytochrome light and dark reactions. Phytochrome dark reactions (i.e. synthesis of P_r and reversion and destruction of P_{fr}) can play a major role in determining the P_{fr}/P_{tot} ratio in dark grown plants at normal physiological temperatures (Heim & Schäfer 1981). In light-grown plants, the physiologically effective phytochrome appears to be relatively stable and phytochrome dark reactions only have a significant effect at very low fluence rates (Jabben & Holmes 1983).

The actinic properties of representative natural daylight and vegetation shade light (which are similar to the spectra in figure 1) are given in table 2. The effect that these changes in spectral quality and quantity have on hypocotyl elongation rate in *C. rubrum* can be derived from studies with artificial light sources that simulate the natural spectra by adding various amounts of far-red light to background white light (Ritter *et al.* 1981).

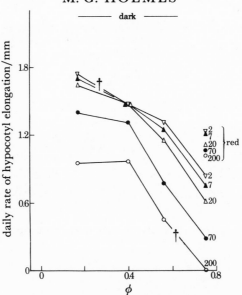

FIGURE 6. The effect of phytochrome photostationary state (ϕ) at a range of phytochrome cycling rates on hypocotyl elongation rate in chlorophyll-free *Chenopodium rubrum* seedlings. The seedlings had been in continuous white light since sowing. The reductions in ϕ were achieved by adding different amounts of far-red light to the background fluorescent white light source. The figure demonstrates that reducing the quantity (cycling rate) of monochromatic red light (red 2:2 min^{-1}, etc.) has a relatively small effect on elongation growth rate compared with reductions in the quantity of polychromatic radiation, irrespective of ϕ (2, 7, etc.: cycling rate of 2 min^{-1}, 7 min^{-1}, etc.). Reductions in ϕ also caused marked increases in growth rate. The combined effects of the reduction in the quantity of blue light, reduction in phytochrome cycling rate, and reduction in ϕ – all of which are characteristic of vegetation shade light compared with natural daylight (from table 2) – result in a large increase in elongation rate (†). The involvement of photosynthesis has been excluded by using chlorophyll-free seedlings. (After Ritter *et al.* (1981).)

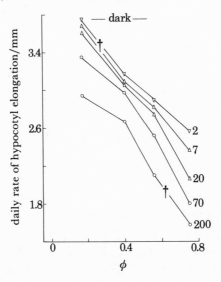

FIGURE 7. The effect of phytochrome photostationary state (ϕ) at a range of phytochrome cycling rates on hypocotyl elongation rate in green *Chenopodium rubrum* seedlings. The seedlings had been in continuous white light since sowing. The reductions in ϕ were achieved by adding different amounts of far-red light to the background white light source. Reducing the quantity of monochromatic red light does not have a consistent effect on hypocotyl growth rate, whereas reducing the quantity of polychromatic radiation causes a marked increase in elongation rate at all ϕ values studied (2, 7, etc.: cycling rate of 2 min^{-1}, 7 min^{-1}, etc.). Reductions in ϕ also cause marked increases in growth rate. The combined effects of the reduction in the quantity of blue light, reduction in phytochrome cycling rate, and reduction in ϕ characteristic of vegetation shade light compared with natural daylight (from table 2) result in a large increase in elongation rate (†). (After Ritter *et al.* (1981).)

The effect of activating the b.a.p. on the response of hypocotyl elongation rate to changes in ϕ is seen in figure 6. The data are plotted for five equal phytochrome cycling rates and the effects of photosynthesis are eliminated by using chlorophyll-free seedlings. Two important points are evident. First, whereas reducing the quantity (indicated by cycling rate) of monochromatic red light (or far-red, not shown) causes only a small increase in elongation rate, a reduction in the quantity of polychromatic radiation causes a relatively large increase in growth rate at all ϕ tested. Second, reducing ϕ at a constant cycling rate also causes an increase in elongation rate; this increase is not caused by the slightly higher level of blue and red light used to maintain constant cycling rate because the same response is observed when blue and red are constant (Holmes & Wagner 1981, 1982; Ritter et al. 1981). Qualitatively similar responses were found in green seedlings (figure 7).

The roles played by the b.a.p. and by phytochrome in the perception of shade by C. rubrum can be interpolated from the curves in figures 6 and 7 by comparing the actinic properties of the experimental light treatments with the actinic properties of daylight and canopy shade light (table 2). It can be seen that the elongation rate of green seedlings (figure 7) in light equivalent to canopy shade († at $\phi = 0.27$ in figure 7) is increased relative to the rate in daylight († at $\phi = 0.61$ in figure 7) for two reasons. First, there is an increase due to the reduction in ϕ. Second, there is an increase due to the reduction in fluence rate (expressed as reduced cycling rate) and this increase is only significant in the presence of blue light. In other words, the reduced photon fluence rate of blue light below vegetation shade relative to daylight is perceived by a b.a.p. and this photoreceptor plays a major role in modulating elongation growth rate in C. rubrum.

(b) The red waveband

(i) Phytochrome as photoreceptor

There is no unequivocal evidence for phytochrome-modulated fluence-rate dependence of elongation growth (i.e. internode) in mature green plants. Although a fluence-rate dependence does exist in green plants in so far as red light affects elongation rate relative to darkness, the relation between the effects of photosynthesis and the effects of phytochrome is unknown.

In light-grown seedlings, phytochrome-modulated hypocotyl growth responses exhibit fluence-rate dependence and the effects caused by photosynthesis and by phytochrome can be separated. The extent of the fluence-rate dependence and the range over which it occurs differs greatly between species. Hypocotyl elongation rate in light-grown Sinapis alba seedlings is strongly dependent on the fluence rate in the red waveband (Wildermann et al. 1978; Beggs et al. 1980). The response cannot be attributed to photosynthesis because chlorophyll-free plants also exhibit a fluence-rate dependence.

There is correlative evidence in dark-grown S. alba seedlings that the fluence-rate dependent component for inhibition of hypocotyl growth by monochromatic red light represents, at least in part, the phytochrome photoconversion processes competing against the phytochrome dark reactions (Heim & Schäfer 1982), and a similar mechanism has been proposed to operate in light-grown plants (Jabben & Holmes 1983). However, at least one factor other than photoconversion must be involved because some fluence-rate dependent response is observed at fluence rates higher than those required for complete photoconversion of P_r to P_{fr} (Beggs et al. 1980; Heim & Schäfer 1982).

Cycling of phytochrome has been proposed as a possible mechanism to explain phytochrome perception of light quantity (see, for example, Jose & Vince-Prue 1978; Johnson & Tasker

1979), although no empirical evidence for this hypothesis has yet been found. In light-grown *S. alba* seedlings, there is no correlation between cycling rate and response (Holmes *et al.* 1982). Morgan *et al.* (1980) used a specific test for the involvement of cycling in regulation of stem elongation in mature *S. alba* plants. They noted that stem extension rate was increased by various wavelengths of far-red light added to a background of white light and that the response was enhanced by increasing the far-red fluence rate. The involvement of cycling *per se* was excluded because addition of monochromatic red light to the white light plus far-red resulted in a decrease, rather than a further increase, in extension rate. Morgan *et al.* (1981) provided further evidence that cycling is not involved in stem extension in mature *S. alba* and that the response to supplementary far-red light is not fluence-rate dependent.

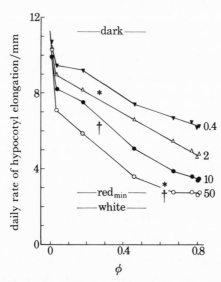

FIGURE 8. The effects of phytochrome photostationary state and photon fluence rate on hypocotyl elongation rate in light-grown *Sinapis alba*: ○, 50 μmol m^{-2} s^{-1}; ●, 10 μmol m^{-2} s^{-1}; △, 2.0 μmol m^{-2} s^{-1}; ▼, 0.4 μmol m^{-2} s^{-1}. The line labelled red$_{min}$ indicates the slowest growth rate (i.e. maximum inhibition) that can be obtained in the red–far-red waveband; the line labelled white indicates the slowest growth rate that can be obtained with white xenon arc light (*ca.* 300 μmol m^{-2} s^{-1} in the 400–800 nm waveband); the small difference between red$_{min}$ and white shows that the b.a.p. makes only a very small contribution to the modulation of elongation growth rate. The figure demonstrates that both a reduction in the red:far-red ratio and a reduction in the quantity of light in the red–far-red waveband contribute to modulation of hypocotyl growth rate. Phytochrome-modulated growth rates are presented for full daylight and canopy light (†, see table 2) and for 10% of daylight and canopy light (*). (After Holmes *et al.* (1982).)

(ii) *Phytochrome in Sinapis alba L.*

Light-grown *S. alba* differs from light-grown *C. rubrum* in that the hypocotyl growth rate shows little response to the quantity of blue light but a marked response to changes in the quantity of red light (Wildermann *et al.* 1978; Beggs *et al.* 1980; Schafer *et al.* 1981). The role played by the quantity of light in the red waveband in the perception of shade can be determined by comparing the actinic properties for phytochrome of daylight and canopy shade light (table 2) with the response of *S. alba* hypocotyls to monochromatic red and far-red light with the equivalent actinic properties (figure 8). Clearly, all wavelengths, or combinations of wavelengths, that produce the same phytochrome photostationary state and cycling rate will be perceived as identical by the plant (Fukshansky *et al.* 1981). For example, a single wavelength that produces a photostationary state of $\phi = 0.30$ is identical for the phytochrome in the plant to a combination of wavelengths (e.g. red plus far-red) that produce the same photostationary state and the same cycling state.

The same phytochrome photoequilibrium established by daylight is produced by monochromatic light at 687 nm (table 2). To obtain the same cycling rate, a photon fluence rate of 452 μmol m^{-2} s^{-1} would be required. The shade light is equivalent for the phytochrome system to monochromatic 701 nm radiation at a photon fluence rate of 17.7 μmol m^{-2} s^{-1}.

The data in figure 8 show that *S. alba* hypocotyl extension growth is controlled by both the quantity and the quality of radiation in the red and far-red wavebands. The photostationary state and cycling rate produced by daylight results in an elongation rate of 2.7 mm per day whereas light equivalent to canopy shade produces an elongation rate of 6.4 mm per day. The important point to note is that decreasing ϕ alone does not produce such a large increase in elongation rate as the combined effects of decreasing ϕ and the quantity of light absorbed by phytochrome. If the fluence rate of both the daylight and canopy light is reduced to 10 %, the role played by the quantity of light is increased.

It must be emphasized that this approach deliberately excludes the small contribution that blue light makes to the response (indicated by the difference between red$_{\min}$ and white in figure 8) and indicates only the two ways by which phytochrome can perceive shade.

5. Conclusions

Evidence has been presented that supports the concept that a fundamental function of phytochrome is to perceive the natural light environment and to modify growth and development accordingly. The photochromic properties of phytochrome enable a plant to respond specifically to changes in light quality caused by surrounding vegetation. This ability to acquire information on the type of shading is of obvious adaptive and survival value as it would allow the plant to react appropriately to competition.

Apart from indicating the extent and nature of shade, phytochrome also functions in some species to detect changes in the fluence rate. Other species appear to be primarily dependent on one or more blue-light-absorbing photoreceptors to detect changes in fluence rate. The perception of changes in fluence rate is important because it enables a plant to respond to both vegetation shade and shading caused by objects that have only a small effect on spectral quality, such as stones and soil. It is noteworthy that the response to fluence rate is immediate and marked in the initial stages of elongation growth (hypocotyl) but plays a lesser role in modulating growth rate in the adult plant (internode). The response to light quality differs in that it is developed only after a few hours of receiving light and that the sensitivity to light quality persists to play the major role in controlling elongation growth in the adult plant.

References

Atkinson, Y. E., Bradbeer, J. W. & Frankland, B. 1980 In *Photoreceptors and plant development* (ed. J. deGreef), pp. 543–550. Antwerpen University Press.

Beggs, C. J., Geile, W., Holmes, M. G., Jabben, M., Jose, A. M. & Schäfer, E. 1981 High irradiance response promotion of a subsequent light induction response in *Sinapis alba* L. *Planta* **151**, 135–140.

Beggs, C. J., Holmes, M. G., Jabben, M. & Schäfer, E. 1980 Action spectra for the inhibition of hypocotyl growth by continuous irradiation in light and dark-grown *Sinapis alba* L. seedlings. *Pl. Physiol.* **66**, 615–618.

Black, M. & Shuttleworth, J. E. 1974 The role of the cotyledons in the photocontrol of hypocotyl extension in *Cucumis sativa* L. *Planta* **117**, 57–66.

Blackman, G. E. & Wilson, G. L. 1951a Physiological and ecological studies in the analysis of plant environment. VI. The constancy for different species of a logarithmic relationship between net assimilation rate and light intensity and its ecological significance. *Ann. Bot.* **15**, 63–94.

Blackman, G. E. & Wilson, G. L. 1951b Physiological and ecological studies in the analysis of plant environment. VII. An analysis of the differential effects of light intensity on the net assimilation rate, leaf-area and relative growth rate of different species. *Ann. Bot.* **15**, 373–408.

Butler, W. L., Lane, H. C. & Siegelman, H. W. 1963 Non-photochemical transformations of phytochrome *in vivo*. *Pl. Physiol.* **38**, 514–519.

Butler, W. L., Siegelman, H. W. & Miller, C. O. 1964 Denaturation of phytochrome. *Biochemistry, Wash.* **3**, 851–857.

Child, R., Morgan, D. C. & Smith, H. 1981 Control of development in *Chenopodium album* L. by shadelight. The effect of light quality (red:far-red ratio) on morphogenesis. *New Phytol.* **89**, 545–555.

Clarkson, D. T. & Hillman, W. S. 1968 Stable concentrations of phytochrome in *Pisum* under continuous illumination with red light. *Pl. Physiol.* **43**, 88–92.

Downs, R. J., Hendricks, S. B., Borthwick, H. A. 1957 Photoreversible control of elongation in Pinto beans and other plants under normal conditions of growth. *Bot. Gaz.* **118**, 199–208.

Evans, L. T., Hendricks, S. B. & Borthwick, H. A. 1965 The role of light in suppressing hypocotyl elongation in lettuce and petunia. *Planta* **64**, 201–218.

Elliott, W. M. 1979 Control of leaf and stem growth in light-grown pea seedlings by two high irradiance responses. *Pl. Physiol.* **63**, 833–836.

Federer, C. A. & Tanner, C. B. 1966 Spectral distribution of light in the forest. *Ecology* **47**, 555–560.

Fitter, A. H. & Ashmore, C. J. 1974 Response of two *Veronica* spp. to a simulated woodland light climate. *New Phytol.* **73**, 997–1001.

Frankland, B. & Letendre, R. J. 1978 Phytochrome and effects of shading on growth of woodland plants. *Photochem. Photobiol.* **27**, 223–230.

Fukshansky, L., Beggs, C. J., Holmes, M. G., Jabben, M. & Schäfer, E. 1981 *Proc. Eur. Symp. Light Mediat. Plant Dev.* no. 9, p. 14.

Grime, J. P. 1966 Shade avoidance and shade tolerance in flowering plants. In *Light as an ecological factor* (ed. R. Bainbridge, G. C. Evans & O. Rackham), pp. 187–207. Oxford: Blackwell.

Grime, J. P. 1981 Plant strategies in shade. In *Plants and the daylight spectrum* (ed. H. Smith), pp. 159–186. London: Academic Press.

Grime, J. P. & Jeffrey, D. W. 1965 Seedling establishment in vertical gradients of sunlight. *J. Ecol.* **53**, 621–642.

Hamdorf, K., Paulsen, R., Schwemer, J. & Tauber, U. 1972 Photoconversion of invertebrate visual pigments. In *Information processing in the visual systems of arthropods* (ed. P. Wehner), pp. 97–108. Berlin: Springer.

Hartmann, K. M. 1966 A general hypothesis to interpret 'high energy phenomena' of photomorphogenesis on the basis of phytochrome. *Photochem. Photobiol.* **5**, 349–366.

Heim, B., Jabben, M. & Schäfer, E. 1981 Phytochrome destruction in dark- and light-grown *Amaranthus candatus* seedlings. *Photochem. Photobiol.* **34**, 89–93.

Heim B. & Schäfer, E. 1982 Light-controlled inhibition of hypocotyl growth in *Sinapis alba* L. seedlings. Fluence rate dependence of hourly light pulses and continuous irradiation. *Planta* **154**, 150–155.

Hendricks, S. B., Borthwick, H. A. & Downs, R. J. 1956 Pigment conversion in the formative responses of plants to radiation. *Proc. natn. Acad. Sci. U.S.A.* **42**, 19–26.

Holmes, M. G. 1975 Studies on the ecological significance of phytochrome. Ph.D. thesis, University of Nottingham.

Holmes, M. G. 1981 Spectral distribution of radiation within plant canopies. In *Plants and the daylight spectrum* (ed. H. Smith), pp. 148–158. London: Academic Press.

Holmes, M. G., Beggs, C. J., Jabben, M. & Schäfer, E. 1982 Hypocotyl growth in *Sinapis alba* L.: the roles of light quality and quantity. *Pl. Cell Environ.* **5**, 45–51.

Holmes, M. G. & Fukshansky, L. 1979 Phytochrome photoequilibria in green leaves under polychromatic radiation: a theoretical approach. *Pl. Cell Environ.* **2**, 59–66.

Holmes, M. G. & McCartney, H. A. 1976 Spectral energy distribution in the natural environment and its implications for phytochrome function. In *Light and plant development* (ed. H. Smith), pp. 467–476. London: Butterworth.

Holmes, M. G. & Smith, H. 1975 The function of phytochrome in plants growing in the natural environment. *Nature, Lond.* **254**, 512–514.

Holmes, M. G. & Smith, H. 1977a The function of phytochrome in the natural environment. I. Characterization of daylight for studies in photomorphogenesis and photoperiodism. *Photochem. Photobiol.* **25**, 533–538.

Holmes, M. G. & Smith, H. 1977b The function of phytochrome in the natural environment. II. The influence of vegetation canopies on the spectral energy distribution of natural daylight. *Photochem. Photobiol.* **25**, 539–545.

Holmes, M. G. & Smith, H. 1977c The function of phytochrome in the natural environment. IV. Light quality and plant development. *Photochem. Photobiol.* **25**, 551–557.

Holmes, M. G. & Wagner, E. 1981 Phytochrome control of hypocotyl extension in light-grown *Chenopodium rubrum* L. *Physiologia Pl.* **53**, 233–238.

Holmes, M. G. & Wagner, E. 1982 The influence of chlorophyll on the spectral control of elongation growth in *Chenopodium rubrum* L. hypocotyls. *Pl. Cell Physiol.* **23**, 745–750.

Jabben, M. 1980 The phytochrome system in light-grown *Zea mays* L. *Planta* **149**, 91–96.

Jabben, M., Beggs, C. & Schäfer, E. 1982 Dependence of P_{fr}/P_{tot} ratios on light quality and light quantity. *Photochem. Photobiol.* **35**, 709–712.

Jabben, M. & Deitzer, G. F. 1978 Spectrophotometric phytochrome measurements in light-grown *Avena sativa* L. *Planta* **143**, 309–313.

Jabben, M., Heim, B. & Schäfer, E. 1980 The phytochrome system in light- and dark-grown dicotyledonous seedlings. in *Photoreceptors and plant development* (ed. J. A. deGreef), pp. 135–158. Antwerpen University Press.

Jabben, M. & Holmes, M. G. 1983 Phytochrome in light-grown plants. In *Encyclopedia of plant physiology*, vol. 16 (*Photomorphogenesis*) (ed. W. Shropshire, Jr & H. Mohr), ch. 27. Berlin: Springer. (In the press.)

Johnson, C. B. 1981 How does phytochrome perceive light quality? In *Plants and the daylight spectrum* (ed. H. Smith), pp. 481–497. London: Academic Press.

Johnson, C. B. & Tasker, R. 1979 A scheme to account quantatively for the action of phytochrome in etiolated and light-grown plants. *Pl. Cell Environ.* **2**, 259–265.

Jose, A. M. & Vince-Prue, D. 1978 Phytochrome action: a reappraisal. *Photochem. Photobiol.* **27**, 209–216.

Kasperbauer, M. J. 1971 Spectral distribution of light in a tobacco canopy and effects of end-of-day light quality on growth and development. *Pl. Physiol.* **47**, 775–778.

Kendrick, R. E. & Spruit, C. J. P. 1972 Light maintains high levels of phytochrome intermediates. *Nature, new Biol.* **237**, 281–282.

Kendrick, R. E. & Spruit, C. J. P. 1973 Phytochrome intermediates *in vivo*. I. Effects of temperature, light intensity, wavelength and oxygen on intermediate accumulation. *Photochem. Photobiol.* **18**, 139–144.

Kendrick, R. E. & Spruit, C. J. P. 1974 Inverse dark reversion of phytochrome: an explanation. *Planta* **120**, 265–272.

Kendrick, R. E. & Spruit, C. J. P. 1976 Intermediates in the photoconversion of phytochrome. In *Light and plant development* (ed. H. Smith), pp. 31–43. London: Butterworth.

Kilsby, C. A. H. & Johnson, C. B. 1982 The *in vivo* spectrophotometric assay of phytochrome in two mature dicotyledonous plants. *Photochem. Photobiol.* **35**, 255–260.

Koukkari, W. L. & Hillman, W. S. 1967 Effects of temperature and aeration on phytochrome transformations in *Pastinaca sativa* root tissue. *Am. J. Bot.* **59**, 1118–1122.

Kumagai, T. 1982 Blue and near ultraviolet reversible photoreaction in the induction of fungal conidiation. *Photochem. Photobiol.* **35**, 123–125.

Lecharny, A. & Jacques, R. 1974 Phytochrome et croissance des tiges; variations de l'effet de la lumière en fonction du temps et du lieu de photoperception. *Physiol. vég.* **12**, 721–738.

Lecharny, A. & Jacques, R. 1979 Phytochrome and internode elongation in *Chenopodium polyspesmum* L. The light fluence rate during the day and the end-of-day effect. *Planta* **146**, 575–577.

Lecharny, A. & Jacques, R. 1980 Light inhibition of internode elongation in green plants. A kinetic study with *Vigna sinensis* L. *Planta* **149**, 384–388.

Löser, G. & Schäfer, E. 1980 Phototropism in *Phycomyces*: a photochromic sensor pigment? In *The blue light syndrome* (ed. H. Senger), pp. 244–250. Berlin: Springer.

McLaren, D. F. & Smith, H. 1978 Phytochrome control of the growth and development of *Rumex obtusifolius* under simulated canopy light environments. *Pl. Cell Environ.* **1**, 61–67.

Meijer, G. 1958 Influence of light on the elongation of gherkin seedlings. *Acta bot. neerl.* **7**, 614–620.

Meijer, G. 1959 Spectral dependence of flowering and elongation. *Acta bot. neerl.* **8**, 189–246.

Mohr, H. 1972 *Lectures on photomorphogenesis*. (237 pages.) New York: Springer.

Mohr, H. 1980 Interaction between blue light and phytochrome in photomorphogenesis. In *The blue light syndrome* (ed. H. Senger), pp. 97–109. Berlin: Springer.

Morgan, D. C., Child, R. & Smith, H. 1981 Absence of fluence rate dependency of phytochrome modulation of stem extension in light-grown *Sinapis alba* L. *Planta* **151**, 497–498.

Morgan, D. C., O'Brien, T. & Smith, H. 1980 Rapid photomodulation of stem extension in *Sinapis alba* L. Studies on kinetics, site of perception and photoreceptor. *Planta* **150**, 95–101.

Morgan, D. C. & Smith, H. 1976 Linear relationship between phytochrome photoequilibrium and growth in plants under simulated natural radiation. *Nature, Lond.* **262**, 210–212.

Morgan, D. C. & Smith, H. 1978*a* The relationship between phytochrome photoequilibrium and development in light grown *Chenopodium album* L. *Planta* **142**, 187–193.

Morgan, D. C. & Smith, H. 1978*b* Simulated sunflecks have large, rapid effects on plant stem extension. *Nature, Lond.* **273**, 534–536.

Morgan, D. C. & Smith, H. 1979 A systematic relationship between phytochrome-controlled development and species habitat, for plants grown in simulated natural radiation. *Planta* **145**, 253–258.

Morgan, D. C. & Smith, H. 1981*a* Control of development in *Chenopodium album* L. by shadelight: the effect of light quantity (total fluence rate) and light quality (red:far-red ratio). *New Phytol.* **88**, 239–248.

Morgan, D. C. & Smith, H. 1981*b* Non-photosynthetic responses to light quality. In *Encyclopedia of plant physiology*, vol. 12a (ed. P. Nobel), pp. 109–134. Springer, Berlin.

Mumford, F. E. & Jenner, E. L. 1971 Catalysis of the phytochrome dark reaction by reducing agents. *Biochemistry, Wash.* **10**, 98–101.

Ogawa, T., Ishikawa, H., Shimada, K. & Shibata, K. 1978 Synergistic action of red and blue light and action spectrum for malate formation in guard cells of *Vicia faba* L. *Planta* **142**, 61–65.

Pratt, L. H. & Briggs, W. R. 1966 Photochemical and nonphotochemical reactions of phytochrome *in vivo*. *Pl. Physiol.* **41**, 467–474.

Pratt, L. H., Shimazaki, Y., Inoue, Y. & Furuya, M. 1982 Analysis of phototransformation intermediates in the pathway from the red-absorbing to the far-red-absorbing form of *Avena* phytochrome by a multichannel transient spectrum analyzer. *Photochem. Photobiol.* **36**, 471–477.

Ritter, A., Wagner, E. & Holmes, M. G. 1981 Light quantity and quality interactions in the control of elongation growth in light-grown *Chenopodium rubrum* L. seedlings. *Planta* **153**, 556–560.

Rombach, J., Bensink, J. & Katsura, N. 1982 Phytochrome in *Pharbitis nil* during and after de-etiolation. *Physiologia Pl.* **56**, 251–258.

Roth-Bejerano, N. 1980 Growth control by phytochrome of de-etiolated bean hypocotyls mediated by chlorophyll fluorescence. *Physiologia Pl.* **50**, 326–330.

Sale, P. J. M. & Vince, D. 1959 Effects of wavelength and time of irradiation on internode length in *Pisum sativum* and *Tropaeoleum majus*. *Nature, Lond.* **183**, 1174–1175.

Satter, R. L. & Wetherell, D. F. 1968 Photomorphogenesis in *Sinningia speciosa*, cv. Queen Victoria. II. Stem elongation: interaction of a phytochrome-controlled process and a red-requiring, energy dependent reaction. *Pl. Physiol.* **43**, 961–967.

Schäfer, E. 1975 A new approach to explain the 'high irradiance responses' of photomorphogenesis on the basis of phytochrome. *J. math. Biol.* **2**, 41–56.

Schäfer, E., Beggs, C. J., Fukshansky, L., Holmes, M. G. & Jabben, M. 1981 A comparative study of the responsivity of *Sinapis alba* L. seedlings to pulsed and continuous irradiation. *Planta* **253**, 258–261.

Schmidt, R. & Mohr, H. 1982 Evidence that a mustard seedling responds to the amount of P_{fr} and not to the P_{fr}/P_{tot} ratio. *Pl. Cell Environ.* **5**, 495–499.

Smith, H. 1975 *Phytochrome and photomorphogenesis*. London: McGraw-Hill.

Smith, H. 1981 a Function, evolution and action of plant photosensors. In *Plants and the daylight spectrum* (ed. H. Smith), pp. 499–508. London: Academic Press.

Smith, H. 1981 b Evidence that P_{fr} is not the active form of phytochrome in light grown maize. *Nature, Lond.* **293**, 163–165.

Smith, H. 1982 Light quality, photoperception and plant strategy. *A. Rev. Pl. Physiol.* **33**, 481–518.

Smith, H. & Holmes, M. G. 1977 The function of phytochrome in the natural environment III. Measurement and calculation of phytochrome photoequilibrium. *Photochem. Photobiol.* **25**, 547–550.

Smith, H. & Morgan, D. C. 1982 The function of phytochrome in nature. In *Encyclopedia of plant physiology*, vol. 16 (*Photomorphogenesis*) (ed. W. Shropshire, Jr & H. Mohr). Berlin: Springer. (In the press.)

Spruit, C. J. P. 1982 Phytochrome intermediates *in vivo*. IV. Kinetics of P_{fr} emergence. *Photochem. Photobiol.* **35**, 117–121.

Stoutjesdijk, P. 1972 a Spectral transmission curves of some types of leaf canopies with a note on seed germination. *Acta bot. neerl.* **21**, 185–191.

Stoutjesdijk, P. 1972 b A note on the spectral transmission of light by tropical rainforest. *Acta bot. neerl.* **21**, 346–350.

Tasker, R. & Smith, H. 1977 The function of phytochrome in the natural environment. V. Seasonal changes in the radiant energy quality in woodlands. *Photochem. Photobiol.* **26**, 487–491.

Thomas, B. 1981 Specific effects of blue light on plant growth and development. In *Plants and the daylight spectrum* (ed. H. Smith), pp. 443–459. London: Academic Press.

Vezina, P. E. & Boulter, D. W. K. 1966 The spectral composition of near UV and visible radiation beneath forest canopies. *Can. J. Bot.* **44**, 1264–1267.

Vince, D. 1956 Studies on the effects of light quality on the growth and development of plants. II. Formative effects in *Lycopersicon esculentum* and *Pisum sativum*. *J. hort. Sci.* **31**, 16–24.

Vince-Prue, D. 1973 Phytochrome and the natural light environment. *Anais Acad. bras. Cienc.* **45** (suppl.), 93–102.

Vince-Prue, D. 1977 Photocontrol of stem elongation in light-grown plants of *Fuchsia hybrida*. *Planta* **133**, 149–156.

Wall, J. K. & Johnson, C. B. 1981 Phytochrome action in light-grown plants: the influence of light quality and fluence rate on extension growth in *Sinapis alba* L. *Planta* **153**, 101–108.

Warrington, I. J., Edge, E. A. & Green, L. M. 1978 Plant growth under high radiant energy fluxes. *Ann. Bot.* **42**, 1305–1313.

Wetherell, D. F. 1969 Phytochrome in cultured wild carrot tissue. I. Synthesis. *Pl. Physiol.* **44**, 1734–1737.

Wildermann, A., Drumm, H., Schäfer, E. & Mohr, H. 1978 Control by light of hypocotyl growth in de-etiolated mustard seedlings. I. Phytochrome as the only photoreceptor pigment. *Planta* **141**, 211–216.

Discussion

A. M. FARMER (*Botany Department, The University, St Andrews, U.K.*). How does Dr Holmes consider that aquatic plants perceive shade? One species of aquatic plant may live in three very different conditions, as regards red and far-red light: (*a*) in deep water shade where the red/far-red ratio is very high; (*b*) in shallow water, but under the canopy of other plants so that it is experiencing green shade, with a low red/far-red ratio; (*c*) in a combination of both green shade and aquatic shade such that the red/far-red ratio is similar to that in the surface irradiance.

However, the resultant shading in each of these conditions may reduce carbon assimilation by a similar degree, and so in this respect the plants are in ecologically comparable situations.

M. G. HOLMES. It is unlikely that phytochrome is responsible for perception of shade caused by water alone in deep aquatic habitats because the changes in red/far-red ratio are not over a range to which phytochrome is very sensitive. In shallow water, however, the variation in red/far-red ratio can be greater (especially where there is an overlying leafy canopy) and phytochrome is ideally suited for the perception of such spectral variations. The increased proportion of far-red light near the surface relative to deep water would be sufficient to induce a relatively large change in phytochrome photoequilibrium and thereby account for morphogenetic changes such as the development of aerial fronds. A blue-light-absorbing photoreceptor would lead to less confusion in both deep and shallow water. It would be naïve to assume that plants, both terrestrial and aquatic, possess only the signal-transducing photoreceptors that we have so far discovered.

M. R. BARTLEY (*Department of Botany, The University, St Andrews, U.K.*). Given that although spectral changes of light penetrating water result in substantial increases in red/far-red ratio with depth, but with little change in the predicted phytochrome photoequilibrium from shallow to deep water, would Dr Holmes consider it possible that the rate of phytochrome cycling (which would decrease dramatically with depth) might play a role in controlling photomorphogenesis underwater, as it does in the photocontrol of seed germination?

M. G. HOLMES. The possibility that phytochrome cycling might play a role in controlling photomorphogenesis cannot be excluded. Unfortunately, no extensive studies have been made in aquatic plants. However, this question has been approached in terrestrial plants. With the exception of photocontrol of seed germination, which Dr Bartley mentioned, the findings so far have provided no empirical evidence for the involvement of phytochrome cycling *per se* in photomorphogenesis.

Phil. Trans. R. Soc. Lond. B **303**, 523–536 (1983)

Printed in Great Britain

The perception of light–dark transitions

By Daphne Vince-Prue

Glasshouse Crops Research Institute, Worthing Road, Littlehampton, Sussex BN16 3PU, U.K.

Transitions between light and darkness are particularly important where these serve as Zeitgebers to synchronize circadian rhythms. A special case is photoperiodism, which depends on the accurate detection of light–dark transitions and on the coupling of this information to a timing mechanism that appears to be based on the circadian clock.

Results from laboratory experiments are considered in relation to the natural changes experienced at dawn and dusk, and evidence is presented that the light–dark transitions that couple to the timing mechanism in short-day plants are perceived through changes in irradiance rather than through changes in light quality.

It has been generally accepted that the *light–dark* transition is sensed by a decrease of P_{fr} levels in darkness, whereas *dark–light* is sensed by the rapid formation of P_{fr} in the light. However, P_{fr} in light-grown plants appears to be rather stable and so changes in P_{fr} level after transfer to darkness may not be a sufficiently accurate method of detecting the light–dark transition in photoperiodism.

The paper reviews some of the evidence from photoperiodic experiments and concludes that the plant may discriminate between light and darkness through the continuous or intermittent formation of 'new' P_{fr}.

Introduction

The ability to discriminate between light and darkness is undoubtedly of great significance in the life of green plants. The initial emergence of the germinating seedlings into the light is accompanied by a profound change in the pattern of development from that characteristic of darkness to that characteristic of the light-grown plant, modulated with respect to the quantity and quality of radiation received. This strategy has presumably evolved to conserve stored reserves until photosynthesis is possible and this *dark–light* transition is primarily sensed by the phototransformation of dark-synthesized P_r into P_{fr}. Having once emerged into the light, plants are then normally subject to daily alternations of light and darkness, and a wide range of behavioural and developmental responses are keyed to this daily cycle.

There are, of course, many direct responses to light, and most of these are probably related to the needs of photosynthesis. A good example of such a response is stomatal opening, which appears to be controlled by light through two distinct systems. One of these operates through photosynthesis, with chlorophyll as its active chromophore. The second uses a photoreceptor located in the guard cells and is sensitive only to blue light (Zeiger *et al.* 1981). In both cases the continued formation of the excited state of the photoreceptor is essential for the response to light, and so discrimination between light and darkness is achieved.

The other main types of response to the daily light–dark cycle are those involving an interaction between a photoreceptor and an endogenous timing system. It is in such responses that transitions between light and darkness achieve their greatest significance because these are the major natural signals to which time measurement is coupled. The phases of overt circadian rhythms such as leaf movements are, for example, entrained to the continuing light–dark cycle in such a way that the response changes with the time of day to the advantage of the plant. There have been relatively few studies of the photoreceptors that perceive the entraining light–dark transitions, but most evidence indicates that phytochrome is a major sensor in higher green plants, although a blue-light absorbing photoreceptor also functions in some cases (Wilkins & Harris 1976; Satter & Galston 1981).

The timing of reproduction and other seasonal events such as dormancy is largely controlled by photoperiodism: a response to the durations of day or night (or both), which change seasonally and so give information about the time of year. The predictive value of photoperiodism is obvious and allows for life cycles to be coordinated to avoid unfavourable conditions such as low temperature in winter, or to be synchronized with favourable ones. Precise timing may be valuable in itself and not just with regard to season, allowing synchronism of events to the advantage of the plant. Synchronized flowering, for example, improves the chances of cross-pollination.

To locate the time of year accurately, time measurement must operate with great precision and be relatively insensitive to random variations in the environment. As might be expected, the greatest precision is often found in plants from the tropics, where the seasonal changes in daylength are much less than those at high latitudes. From observations made in West Africa, it was found that a difference of only 15 min of light per day was sufficient to determine whether or not flowering occurred in several species of tropical short-day plant (Njoku 1958). Such high precision in time measurement requires similarly high precision in the perception of the transitions between light and darkness. Because their flowering often occurs most rapidly in continuous light, the significance of light–dark transitions is less clear for long-day plants than for short-day plants, where flowering is primarily dependent on the duration of darkness (Vince-Prue 1979, 1981). Consequently the mechanism through which light–dark transitions are sensed is discussed in this paper in relation to the perception of the signals that synchronize photoperiodic time measurement in short-day plants. The discussion is largely based on results obtained with *Pharbitis nil*.

CHARACTERISTICS OF THE SHORT-DAY RESPONSE

Perhaps the best known characteristic of photoperiodism in short-day plants is that their photoperiodic time-keeping is essentially a question of measuring the duration of uninterrupted darkness, which must exceed a certain critical value to allow flowering to occur (Vince-Prue 1975). Because the overall duration of darkness is critical for flowering, both the beginning (dusk) and end (dawn) of the night must be sensed precisely. A second important characteristic is that an inductive dark period can be rendered ineffective by an interruption with a short light treatment (night-break) given at a particular time. In experiments with *Pharbitis nil*, this light-sensitive period occurs 8–9 h after the transition to darkness irrespective of its duration, and similar results have been obtained with *Xanthium strumarium*. The night-break response is thus a transient period of light sensitivity related in time to the beginning rather than the end of the dark period. Although the time of night-break sensitivity may more or less coincide with the critical night length, it always occurs somewhat earlier (Takimoto & Hamner 1964) and, under some conditions, appreciably so (figure 1). After the transient light-sensitive phase is over, therefore, further reactions are necessary for the induction of flowering; these reactions are also inhibited by light. They are also temperature-sensitive in *Pharbitis* so that the critical night length, but not the time of night-break sensitivity, is influenced by temperature. However, when repeated cycles are used, as under natural conditions, the critical dark period is hardly influenced by temperature (Takimoto 1969).

The capacity to respond to an inductive dark period can be modified and in some cases eliminated by appropriate manipulation of the preceding photoperiod. In single-cycle short-day

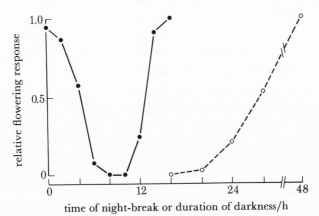

FIGURE 1. Time of sensitivity to a night-break and critical length of night in *Pharbitis nil*. Plants received a single dark period of various durations (o) or a 48 h dark period interrupted at different times with a 5 min red night-break (●). The temperature of the dark period was 18 °C in both cases. (From Takimoto & Hamner (1964).)

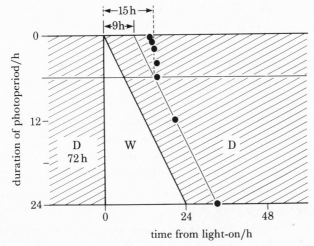

FIGURE 2. Time of maximum sensitivity to a red night-break as a function of the duration of the preceding photoperiod. The time of maximum sensitivity is indicated in real time (●) for the first circadian cycle only. (From Lumsden *et al.* (1982).)

plants, such as *Pharbitis*, where exposure to a single dark period is sufficient for flowering, there is apparently no maximum duration of photoperiod beyond which a subsequent dark period longer than the critical ceases to induce flowering. When the duration of light is decreased, however, there is a change in the timing characteristics and, when the photoperiod is 5 h or less, timing of night-break sensitivity is coupled to the beginning (15 h after light-on signal) rather than to the end (9 h after light-off signal) of the day (figure 2).

Time measurement can thus be coupled either to a light-on or to a light-off signal, the latter being perceived only after exposure to several hours of continuous light. Recent experiments have shown that the critical night length is timed in the same way, i.e. from the beginning or the end of the photoperiod depending on its duration (P. J. Lumsden, unpublished). A very similar pattern of response has been reported for another single-cycle short-day plant, *Xanthium strumarium* (Papenfuss & Salisbury 1967).

In *Pharbitis*, a skeleton photoperiod beginning and ending with a pulse of red light is sufficient to induce flowering in response to a subsequent dark period (Friend 1975). The time of sensitivity to a night-break remains characteristic of a single pulse rather than continuous light and occurs 15 h after light-on (from the second pulse) (Lumsden *et al.* 1982). It has been shown that P_{fr} is required for the flowering response under these conditions, and the reversibility of the first pulse by far-red light given 12 h later indicates the stability of the P_{fr} formed by the first pulse (Friend 1975). However, recent experiments (H. Saji, unpublished) have indicated that, with respect to night-break timing, a 24 h skeleton is perceived as continuous light (i.e. the night-break sensitivity occurs 9 h after light-off) when brief pulses of red light are given every hour. Thus, although a relatively stable P_{fr} is involved in the induction of flowering, exposure for several hours to continuous light or to frequent pulses of red light is necessary for timing to be coupled to a light-off signal.

WHAT SIGNALS ARE RESPONDED TO UNDER NATURAL CONDITIONS?

Under natural conditions the transitions between light and darkness occur through a gradually changing irradiance, especially at high latitudes: these are accompanied by a change in spectral quality (figure 3). Twilight spectra are relatively rich in blue and far-red wavelengths and relatively poor in orange–red light (Holmes & McCartney 1976; Smith & Morgan 1981). The pattern of spectral distribution becomes more exaggerated as the solar elevation declines and, during evening twilight on cloudless days, the ratio of red to far-red light decreases from a daylight value of about 1.1 to values in the region of 0.8 to 0.7 (Smith 1982). Recently reported higher values for daylight and twilight (Salisbury 1981) are due mainly to the chosen bandwidths and to the presentation of the data in terms of energy rather than quantum ratios; nevertheless the trend towards a reduced red:far-red ratio during twilight is clearly evident (figure 3). Either the decrease in irradiance or the lowering of the red:far-red ratio during evening twilight could therefore be the transition signal for the beginning of dark-time measurement at dusk. The opposite change could signal dawn. The ratio of either blue:green, or blue:red could also provide a reliable index of the progression through twilight (Smith 1982).

When does a plant begin to respond to darkness in the evening? Or to light at dawn? Relatively few experiments have considered these questions under natural conditions. In one such experiment in Japan, plants were transferred to darkness at different times during evening twilight and returned to daylight at different times in the morning, to determine when dark-time measurement began and ended (Takimoto & Ikeda 1961). Under these conditions, the inductive dark period for *Pharbitis nil* began when the natural irradiance had fallen to between $1-2$ W m^{-2} in the evening and ended when it had increased to only about 0.004 W m^{-2} in the morning (Takimoto & Ikeda 1961). Thus, for *Pharbitis*, the biological night began near the time of astronomical sunset and ended at about the beginning of civil twilight. The responses of several other short-day species were also examined and showed considerable variation in the irradiance values at which the night began and ended. In another experiment in Hawaii, a twilight irradiance below about 0.15 W m^{-2} was found to be equivalent to photoperiodic darkness for sugar cane (Clements 1968). Although tropical twilights are relatively short, the twilight time during which the irradiance exceeded 0.15 W m^{-2} increased the effective daylength by some 26 min compared with the times of sunrise and sunset.

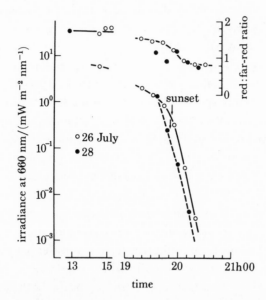

FIGURE 3. Changes with time in irradiance and in red:far-red ratio during evening twilight. Measurements were made on two separate days in Logan, Utah. (From Salisbury (1981).)

As changes in the red:far-red ratio were occurring together with irradiance during these natural twilight periods, this experimental approach does not rule out the possibility that irradiance was being sensed through the perception of an accompanying change in light quality, as has been proposed for the detection of canopy shade by phytochrome (Smith 1982). Smith has pointed out that the changes in red:far-red ratio during twilight are in fact very variable because of cloud conditions on the horizon, and so may not be precise signals for dusk perception. However, this is also true of twilight irradiance, which can vary considerably from day to day because of clouds. Such variations could lead to significant errors in timing, and indeed the effective photoperiod for *Pharbitis* has been shown to be 20–30 min longer on clear days than on cloudy ones (Takimoto & Ikeda 1960). The error would become very small, however, for plants such as *Xanthium* with low threshold irradiances (Salisbury 1963). It could also be reduced, by averaging, when repeated cycles are necessary for induction.

Changes in the quality and quantity of twilight occur together, and both can vary with cloud cover, latitude and season. Consequently experimental approaches with artificial light sources that vary each factor independently are necessary to determine whether either the quality or quantity of light, or both, are important natural cues for the perception of light–dark transitions in photoperiodism.

CHANGES IN END-OF-DAY LIGHT QUALITY

Most experimental approaches to end-of-day light quality have considered only its effects on the magnitude of the overall flowering response. It is abundantly clear from the results of many experiments that exposing plants to far-red light before entry to darkness can substantially reduce or even eliminate flowering (Vince-Prue 1975, 1981) (figure 5). However, these conditions are far from natural, involving, as they usually do, either very short photoperiods or a single very long dark period. With longer photoperiods under natural 24 h cycles, an end-of-day light treatment with far red has been shown to promote flowering in some cases (Cumming 1963; Fredericq 1964; Esashi & Oda 1964) and so the possibility cannot be excluded that the low red:far-red ratio that occurs during evening twilight increases the flowering response when daylengths are close to the critical. However, as far as I am aware, this possibility has never been systematically tested under natural conditions.

These effects of end-of-day light quality on the flowering response do not, however, provide any information about the nature of the signal that initiates dark-time measurement, and it is necessary to design experiments specifically to determine how timing is affected. When the time of maximum sensitivity to a night-break was used as an indicator, an end-of-day treatment with 5 min of far red appeared to have almost no effect on timing in *Pharbitis* (Vince-Prue 1981; Takimoto & Hamner 1965a). Similarly, there was little or no effect on critical night length in *Chenopodium* (King & Cumming 1972) or *Xanthium* (Salisbury 1981) when the photoperiod was terminated with far-red or far-red-enriched light.

At most, therefore, there seems to be only a very small change in timing when the red:far-red ratio is reduced at the end of the day much more drastically than would ever occur under natural conditions. It must be emphasized, however, that these experiments have been carried out on only a limited range of species and usually with a single inductive cycle. It is necessary to investigate a wider range of plants and conditions, using both the time of night-break sensitivity and the critical night length as probes of time measurement to determine whether this conclusion is generally justified.

CHANGES IN END-OF-DAY IRRADIANCE

Varying the irradiance without changing the light quality has shown clearly that dark time measurement begins when the irradiance is below a critical value. In *Xanthium*, time measurement proceeded normally when plants were transferred to white light of low irradiance (*ca.* 100 mW m^{-2}) (Salisbury & Ross 1969), but in this experiment the change in light quantity was accompanied by some change in quality from daylight to incandescent. More convincing are the results of Takimoto (1967) for *Pharbitis*, where plants remained in fluorescent light throughout. Under these conditions time measurement began when plants were transferred to 40 mW m^{-2} but was completely prevented by light at *ca.* 1.0 W m^{-2} or more (figure 4). In both cases, night-breaks were used as a probe of time measurement. In an attempt to approximate more closely to natural conditions, we have recently looked at the effect of transferring plants through several different stepped gradients of irradiance without changing light quality (table 1). In all cases dark time measurement began when plants were transferred to *ca.* 2 W m^{-2} (continuous white fluorescent light) and a *change* in irradiance *per se* was unimportant. Although more experiments are needed with other plant species, there seems little doubt that dark time measurement can begin when the irradiance decreases below a threshold value in the absence of any change in light quality. Fewer studies have been made on the dawn signal, but the limited evidence indicates that here too a change in light quantity can terminate the dark period.

The fact that, under natural conditions, the irradiance fell from that perceived by *Xanthium* plants as light to that perceived as darkness within 5–12 min on two successive days (Salisbury 1981) demonstrates that the decrease in irradiance during evening twilight can give a precise environmental cue for photoperiodic timing.

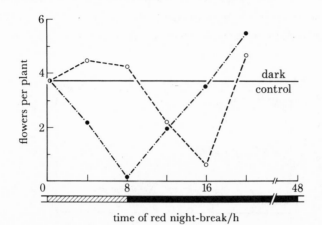

time of red night-break/h

FIGURE 4. Effect of irradiance of an 8 h extension on the initiation of dark-time measurement in *Pharbitis nil*. After exposure to continuous white fluorescent light (16 W m^{-2}) plants received an 8 h extension with fluorescent light at 40 m W m^{-2} (●) or 1 W m^{-2} (○) before transfer to a 40 h inductive dark period. Time measurement was examined by exposing plants to a 15 min red night-break at various times. (From Takimoto (1967).)

TABLE 1. EFFECT OF STEPPED GRADIENTS OF IRRADIANCE AT THE END OF DAY ON DARK TIME MEASUREMENT IN *PHARBITIS NIL*

(The time of maximum sensitivity to a night-break normally occurs 9 h after transfer to darkness. All plants received a 32 h photoperiod followed by a 48 h inductive dark period. (Data of P. J. Lumsden.))

24 h	light treatment/(W m^{-2}) 2 h	2 h	2 h	2 h	time of maximum night-break sensitivity from end of day/h
80	60	40	20	2	7
40	30	20	10	2	7
20	15	10	5	2	7

PHOTORECEPTORS FOR PHOTOPERIODISM

Critical studies of the photoreceptor(s) for photoperiodism began with the discovery of the effect of a brief night-break that allowed the construction of action spectra for the control of flowering in short-day plants (Parker *et al.* 1946). These, together with demonstrations of red–far-red reversibility (Downs 1956) have shown that night-breaks are perceived through the formation of P_{fr}. Based on these results and on the observations of the disappearance of P_{fr} after transfer to darkness, it has generally been accepted that the dawn signal is sensed by the photochemical formation of P_{fr} and dusk by the lowering of P_{fr} levels through non-photochemical reactions (Vince-Prue 1982).

The assumption that phytochrome is the photoreceptor that discriminates between light and darkness in short-day plants largely rests on two kinds of evidence: (1) that the inhibition of flowering by a night-break operates through the formation of P_{fr} and (2) that P_{fr} is required during and sometimes after the photoperiod for the induction of flowering (Vince-Prue 1975).

[183]

However, neither of these is unequivocal evidence for a role for P_{fr} in the perception of the light–dark transitions that begin and end the critical night. The suppression of flowering by end-of-day far red (figure 5) and its reversal by red (Takimoto & Hamner 1965 *b*) shows only that P_{fr} is required for flowering. Although the night-break is often assumed to be equivalent to dawn, sensitivity to a night-break often occurs before the end of the critical night (figure 1), so that the night-break and dawn may not necessarily be sensed in the same way.

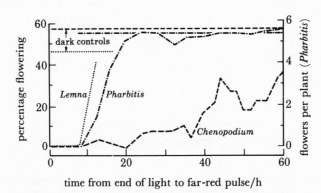

time from end of light to far-red pulse/h

FIGURE 5. Effect on flowering of the time of giving a far-red light pulse during a single inductive dark period. Data for *Lemna paucicostata* 441 from Saji *et al.* (1982) and for *Chenopodium rubrum* from Cumming *et al.* (1965); plants were previously in continuous white light. *Pharbitis nil* was previously exposed to a single photoperiod of 23 h white light (data of P. J. Lumsden).

There is, however, no good evidence for the participation of photoreceptors other than phytochrome in the photoperiodic control of flowering. Photosynthesis, although undoubtedly necessary for the expression of flowering, does not appear to be directly involved in the perception of photoperiodic signals. In *Pharbitis*, for example, normal photoperiodic induction and time measurement have been shown to occur in photobleached seedlings in the absence of photosynthesis (King & Vince-Prue 1978). An essential contribution from a blue-absorbing photoreceptor has not been demonstrated and most of the reported effects of blue light can be explained on the assumption that phytochrome is the photoreceptor (Vince-Prue 1979). When given as a 24 h photoperiod, low-irradiance red light is just as effective as high-irradiance white light for photoperiodic induction in *Pharbitis*, whereas blue light is much less effective even when terminated by red (Vince-Prue 1981). Moreover, when given as an extension to a photoperiod in white light, a much lower irradiance of red light, compared with blue, is effective in preventing the initiation of dark time measurement (Takimoto 1967).

In *Pharbitis*, there is good evidence that the 'light-on' signal is perceived by phytochrome. A single 5 min pulse of red light is sufficient to initiate a circadian rhythm of sensitivity to a night-break and leads to the induction of flowering in response to a dark period longer than critical (King *et al.* 1982; Ogawa & King 1979). Although reversibility by far red has not been tested with respect to the initiation of timing (and may not be possible because of the requirement for P_{fr} for induction), the effectiveness of such a brief exposure to red light strongly suggests that phytochrome is the photoreceptor. The greater effectiveness of red light in preventing the beginning of dark timing when given as an extension of photoperiod (Takimoto 1967; Salisbury 1981) indicates that phytochrome is also the photoreceptor for the photoperiodic light-off signal.

[184]

If perception of a light–dark transition is to be accounted for solely in terms of P_{fr}, the precision of coupling to dark time measurement and the fact that 'darkness' does not begin appreciably sooner when P_{fr} is drastically lowered photochemically at the beginning of the night require that P_{fr} levels must fall rapidly through non-photochemical reactions. (An early proposal that non-photochemical loss of P_{fr} is itself the basic timer (Hendricks 1960) has been excluded by a variety of approaches (cf. Vince-Prue 1975).) Since most of the available evidence indicates that a decrease in irradiance below a threshold level constitutes the light-off signal, it is also necessary to propose that the loss of P_{fr} through non-photochemical reactions becomes significantly large at low irradiances and so allows dark time measurement to begin.

One problem with this hypothesis is that part of the total P_{fr} pool appears to be rather stable in darkness. Because most of the phytochrome has already decayed in fully de-etiolated plants, P_{fr} is present almost exclusively in this stable pool and the reduction in P_{fr} level after transfer to darkness would be small and very slow (Heim *et al.* 1981). Consequently non-photochemical changes in P_{fr} level may not be a sufficiently accurate method of detecting the light–dark transition that initiates dark time measurement in light-grown plants.

Much of the early evidence for stable P_{fr} came from physiological experiments in which far-red pulses were given at different times after transfer to darkness. In *Pharbitis*, the apparent stability of P_{fr} can easily be demonstrated and far-red pulses can be shown to depress flowering even after many hours in darkness (figure 5). These and similar results obtained in other experiments and with other plants (figure 5) are not consistent with the concept that the light–dark transition is sensed through a rapid non-photochemical lowering of P_{fr} unless there are kinetically separate pools of phytochrome with different physiological functions: an unstable pool of P_{fr} coupled to time measurement, and a stable pool required for the induction of flowering.

Spectrophotometric studies

The existence of two independent pools of P_{fr} has recently been suggested on the basis of spectrophotometric data from *Amaranthus caudatus* seedlings (Brockmann & Schäfer 1982). Moreover, measurements made on norfluorazon-treated seedlings of *Pharbitis nil* have shown that there was still some P_{fr} present in the unstable pool even after 3–4 days in continuous white light; this unstable P_{fr} had essentially disappeared 30–60 min after transfer to darkness (figure 5 in Heim *et al.* 1981). These results are very similar to earlier measurements made on *Pharbitis* cotyledons photobleached by exposing them to high-irradiance light for 24 h at low temperature (Vince-Prue *et al.* 1978). In these cotyledons, P_{fr} fell below the limit of detection within 30–40 min after the transfer to darkness, but sensitivity was too poor to detect the continued presence of P_{fr} in the stable pool.

More recently, detailed studies have been made on *Pharbitis* cotyledons that had either been photobleached at low temperature or after treatment with norfluorazon (Rombach *et al.* 1982). After exposure to continuous light for two 12 h periods, such as might be experienced under natural daylengths, evidence was found for both stable and unstable components of the total P_{fr}. There were also indications that P_{fr} could be lost both through reversion to P_r and by destruction to a non-photoreversible form. A rapidly reverting pool of P_{fr} was not observed in etiolated plants and the capacity for dark reversion appeared to develop only after exposure to light for several hours; it also disappeared in darkness. In this context it is interesting to note again (figure 2) that a light-off signal was not perceived in *Pharbitis* until the plants had been exposed to continuous light for more than 6 h.

The spectrophotometric evidence therefore does not exclude the possibility that there is a distinct pool of P_{fr} that undergoes rapid loss through non-photochemical reactions and so could be involved in the perception of the light–dark transition that couples to photoperiodic time measurement. Some of the P_{fr} is lost rapidly through non-photochemical reactions even in light-grown plants with $t_{\frac{1}{2}}$ values sufficiently short to give a precise transition signal (Hillman 1964; Heim *et al.* 1981; Vince-Prue *et al.* 1978; Rombach *et al.* 1982).

Physiological studies

A physiological approach to understanding how the light–dark transition might be perceived is to determine what experimental treatments act in the same way as continuous light and prevent the beginning of dark time measurement (Takimoto 1967; Salisbury 1981). In recent experiments with *Pharbitis*, we have routinely used a single 24 h photoperiod under continuous white fluorescent light at 80 W m^{-2} followed by a single 48 h dark period at 25 °C; plants are then transferred to continuous white fluorescent light. This treatment results in a high level of flowering, which is strongly inhibited when a red night-break is given at a particular time in the inductive dark period (Lumsden *et al.* 1982). The time of maximum sensitivity to a night-break is sharply defined at about 9 h after the light-off signal. When this 24 h photoperiod is extended for 6 h with continuous light, a delay of 6 h in the time of maximum night-break sensitivity would be expected, and the effectiveness of different kinds of extension treatments can therefore be evaluated. The 6 h extension period was always terminated with a 5 min exposure to red, to establish the same $P_{fr}:P_{total}$ ratio before transfer to darkness (figure 6).

As expected, an extension with 6 h of darkness terminated by 5 min red had little effect, and timing was essentially coupled to the initial transfer to darkness (figure 6). Continuous red light, on the other hand, effectively delayed the onset of dark timing, which was coupled to the end of the red extension. The extension light did not, however, have to be continuous, and complete delay occurred when red pulses lasting 1 or 5 min were given at hourly intervals; red pulses given at 2 or 3 h intervals gave a partial delay but were not equivalent to continuous light. These results are consistent with the concept that P_{fr} is lost rapidly in darkness and must be maintained by frequent pulses if the treatment is to be perceived as continuous light. A similar delay in dark time measurement after an extension with half-hourly pulses of red has also been observed in the opening of *Pharbitis* flowers (Kaihara & Takimoto 1980).

When the reversibility of the red pulses by far-red was examined, however, the results did not support the hypothesis that repeated exposures to light are necessary because P_{fr} is lost in the intervening dark periods. We have found no evidence for reversibility of the red pulses by far red (figure 6), as would be expected if this were so. Control treatments with far-red pulses given alone had much less effect than red, showing that it should have been possible to demonstrate reversibility under these conditions. Coupling of P_{fr} to the time-measuring system must therefore occur very rapidly, or alternatively the action of red pulses is not through P_{fr} formation.

One possibility consistent with these results is the concept of new P_{fr}, which was developed to explain somewhat similar observations in *Mesotaenium*. Repeated exposure to red every few minutes was needed to maintain chloroplast movement, even though P_{fr} was shown to be still present (Haupt & Reif 1979). The authors suggested that P_{fr} 'ages' in some way and is then no longer effective physiologically. In *Pharbitis*, a separate pool of phytochrome with a different physiological function would also have to be present because far-red pulses have demonstrated that P_{fr} continues to affect the flowering response for many hours (figure 5).

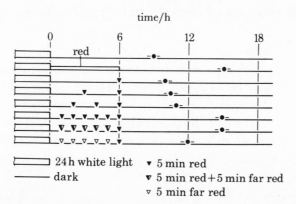

FIGURE 6. Effect of day-extension treatments on time of maximum sensitivity to a night-break for the control of flowering in *Pharbitis nil*. Plants were exposed to a single photoperiod of 24 h white light followed by a day extension for 6 h, as indicated on the figure, before transfer to a 48 h inductive dark period. Time measurement was examined by exposing plants to a 10 min red night-break at different times. The approximate time at which the night-break effect was greatest is indicated (–●–). Red (day-extension pulses and night-breaks) 55 µmol m^{-2} s^{-1}; far-red (700–780 nm) 9 µmol m^{-2}; continuous red, 12 µmol m^{-2}. (Data of P. J. Lumsden.)

Because the requirement for repeated exposure to light is not simply a function of the rapid loss of P_{fr} in darkness, the hypothesis that, with respect to the perception of light–dark transitions in photoperiodism, the plant senses 'continuous' light through the continuous or intermittent formation of new P_{fr} seems the most consistent with our experimental data at present.

Conclusions

Evidence has been presented that the light–dark transition that couples to the initiation of dark time measurement is perceived through a reduction in irradiance and that the photosensory pigment is phytochrome. The precise way in which this occurs is still uncertain but there is some evidence to suggest that this cannot be accounted for on the basis simply of the amount of P_{fr} that is present, as seems to be true of many other physiological responses under phytochrome control.

The general properties of phytochrome do not at first sight appear to make it a particularly good candidate for the precise sensing of light–dark transitions and for the function of discriminating between light and darkness. A better candidate would be a pigment that sensitizes reactions from the excited state. The characteristic photoreversibility of phytochrome, so appropriate for the detection of canopy shade through changes in light quality, appears to be of little consequence in the detection of the light–dark signals for photoperiodic time measurement. Consequently, light–dark sensing has usually been considered a function of the non-photochemical reactions that lead to a reduction in the amount of P_{fr} present in the tissue after transfer to darkness or to a sufficiently low irradiance. The photochemical formation of P_{fr} would then signal dawn.

However, experimental evidence for the stability of P_{fr}, particularly in light-grown plants, suggests that these non-photochemical reactions may not be sufficiently rapid for the accurate sensing of the time of transfer to darkness. Another problem is the relative lack of sensitivity to changes in temperature (Vince-Prue 1975). A highly unstable pool of phytochrome with a specific function for the perception of photoperiodic Zeitgebers remains a possibility, and there is spectrophotometric evidence for an unstable pool of P_{fr} in both light-grown and dark-grown plants.

Physiological approaches using intermittent exposure to red light support the concept that P_{fr} may be lost rapidly in the intervening dark periods. However, the lack of reversibility by far-red argues against the conclusion that repeated exposures are required because of the rapid non-photochemical loss of P_{fr}. A concept consistent with these results is that P_{fr} ages in some way so that the perception of light as being continuous occurs only when new P_{fr} is regenerated sufficiently rapidly. If this is so, then phytochrome appears to operate somewhat differently in the detection of the continuous light signal for photoperiodic timing than in, for example, the perception of canopy shade.

Photoperiodism has been observed in an enormous range of organisms throughout the plant and animal world, and these organisms have adopted a variety of photoreceptors for the perception of photoperiodic Zeitgebers. Following the premise of Smith (1982), these photoreceptors must have been selected for their ability to acquire specific information about the light environment, namely to detect precisely the transitions between light and darkness that are coupled to photoperiodic time measurement. In higher plants, the photoperiodic photoreceptor appears to have more than one function: to detect light quality (responses to shade (Smith 1982)), to detect light direction (chloroplast orientation (Haupt 1982)) and to discriminate between light and darkness (photoperiodism). Thus phytochrome is able to detect different properties of light and it is not clear which , if any, of these can be identified as the fundamental perceptual function of the photoreceptor. Perhaps an original function has been modified and a different perceptual mechanism may have evolved to allow accurate sensing of the transitions between light and darkness in photoperiodism.

A final word must concern the blue-absorbing photoreceptor that can operate as a very sensitive detector of light quantity changes during twilight for the control of stomatal opening (Zeiger et al. 1981). It is interesting to speculate why this photoreceptor does not appear to have been adopted for plant photoperiodism.

I thank P. J. Lumsden for permission to quote from his unpublished results.

References

Brockmann, I. & Schäfer, E. 1982 Analysis of P_{fr} destruction in *Amaranthus caudatus* L.: evidence for two pools of phytochrome. *Photochem. Photobiol.* **35**, 555–558.

Clements, H. F. 1968 Lengthening versus shortening dark periods and blossoming in sugar cane as affected by temperature. *Pl. Physiol.* **43**, 57–60.

Cumming, B. G. 1963 Evidence of a requirement for phytochrome P_{fr} in the floral induction of *Chenopodium rubrum*. *Can. J. Bot.* **41**, 901–926.

Cumming, B. G., Hendricks, S. B. & Borthwick, H. A. 1965 Rhythmic flowering responses and phytochrome changes in a selection of *Chenopodium rubrum*. *Can. J. Bot.* **43**, 825–853.

Downs, R. J. 1956 Reversibility of flower initiation. *Pl. Physiol.* **31**, 279–284.

Esashi, Y. & Oda, Y. 1964 Inhibitory effect of far-red light on the flowering of *Xanthium pensylvanicum*. *Pl. Cell Physiol.* **5**, 507–511.

Fredericq, H. 1964 Conditions determining effects of far-red and red irradiations on the flowering response of *Pharbitis nil. Pl. Physiol.* **39**, 812–816.

Friend, D. J. C. 1975 Light requirements for photoperiodic sensitivity in cotyledons of dark-grown *Pharbitis nil. Physiologia Pl.* **35**, 286–296.

Haupt, W. 1982 Light-mediated movement of chloroplasts. *A. Rev. Pl. Physiol.* **33**, 205–233.

Haupt, W. & Reif, G. 1979 'Ageing' of phytochrome P_{fr} in *Mesotaenium. Z. PflPhysiol.* **92**, 153–169.

Heim, B., Jabben, M. & Schäfer, E. 1981 Phytochrome destruction in dark- and light-grown *Amaranthus caudatus* seedlings. *Photochem. Photobiol.* **34**, 89–93.

Hendricks, S. B. 1960 Rate of change of phytochrome as an essential factor determining photoperiodism. *Cold Spring Harb. Symp. quant. Biol.* **25**, 245–248.

Hillman, W. S. 1964 Phytochrome levels detectable by *in vivo* spectrophotometry in plant parts grown or stored in the light. *Am. J. Bot.* **51**, 1102–1107.

Holmes, M. G. & McCartney, H. A. 1976 Spectral energy distribution in the natural environment and its implications for phytochrome function. In *Light and plant development* (ed. H. Smith), pp. 467–476. London: Butterworths.

Kaihara, S. & Takimoto, A. 1980 Studies on the light controlling the time of flower opening in *Pharbitis nil. Pl. Cell Physiol.* **21**, 21–26.

King, R. W. & Cumming, B. G. 1972 The role of phytochrome in photoperiodic time measurement and its relation to rhythmic timekeeping in the control of flowering in *Chenopodium rubrum. Planta* **108**, 39–57.

King, R. W., Schafer, E., Thomas, B. & Vince-Prue, D. 1982 Photoperiodism and rhythmic response to light. *Pl. Cell Environ.* **5**, 395–404.

King, R. W. & Vince-Prue, D. 1978 Light requirement, phytochrome and photoperiodic induction of flowering of *Pharbitis nil* Chois. I. No correlation between photomorphogenetic and photoperiodic effects of light pretreatment. *Planta* **141**, 1–7.

Lumsden, P., Thomas, B. & Vince-Prue, D. 1982 Photoperiodic control of flowering in dark-grown seedlings of *Pharbitis nil* Choisy. *Pl. Physiol.* **70**, 277–282.

Njoku, E. 1958 The photoperiodic response of some Nigerian plants. *Jl W. Afr. scient. Ass.* **4**, 99–111.

Ogawa, Y. & King, R. W. 1979 Establishment of photoperiodic sensitivity by benzyladenine and a brief red irradiation in dark-grown seedlings of *Pharbitis nil* Chois. *Pl. Cell Physiol.* **20**, 115–122.

Papenfuss, H. D. & Salisbury, F. B. 1967 Properties of clock re-setting in flowering of *Xanthium. Pl. Physiol.* **42**, 1562–1568.

Parker, M. W., Hendricks, S. B., Borthwick, H. A. & Scully, N. J. 1946 Action spectrum for the photoperiodic control of floral initiation of short-day plants. *Bot. Gaz.* **108**, 1–26.

Rombach, J., Bensink, J. & Katsura, N. 1982 Phytochrome in *Pharbitis nil* during and after de-etiolation. *Physiologia Pl.* **56**, 251–258.

Saji, H., Furuya, M. & Takimoto, A. 1982 Spectral dependence of night-break effect on photoperiodic floral induction in *Lemna paucicostata* 441. *Pl. Cell Physiol.* **23**, 623–629.

Salisbury, F. B. 1963 *The flowering process.* Oxford: Pergamon Press.

Salisbury, F. B. 1981 Twilight effect: initiating dark time measurement in photoperiodism of *Xanthium. Pl. Physiol.* **67**, 1230–1238.

Salisbury, F. B. & Ross, C. 1969 *Plant physiology*, p. 747. Belmont: Wadsworth Publishing Co.

Satter, R. L. & Galston, A. W. 1981 Mechanisms of control of leaf movements. *A. Rev. Pl. Physiol.* **32**, 83–110.

Smith, H. 1982 Light quality, photoperception, and plant strategy. *A. Rev. Pl. Physiol.* **33**, 481–518.

Smith, H. & Morgan, D. 1981 The spectral characteristics of the visible radiation incident upon the surface of the earth. In *Plants and the daylight spectrum* (ed. H. Smith), pp. 5–20. London: Academic Press.

Takimoto, A. 1967 Studies on the light affecting the initiation of endogenous rhythms concerned with the photoperiodic responses in *Pharbitis nil. Bot. Mag., Tokyo* **80**, 241–247.

Takimoto, A. 1969 *Pharbitis nil* Chois. In *The induction of flowering* (ed. L. T. Evans), pp. 90–115. Melbourne: Macmillan of Australia.

Takimoto, A. & Hamner, K. C. 1964 Effect of temperature and preconditioning on photoperiodic response of *Pharbitis nil. Pl. Physiol.* **39**, 1024–1030.

Takimoto, A. & Hamner, K. C. 1965a Effect of far-red light and its interaction with red light in the photoperiodic response in *Pharbitis nil. Pl. Physiol.* **40**, 859–864.

Takimoto, A. & Hamner, K. C. 1965b Kinetic studies on pigment systems concerned with the photoperiodic response in *Pharbitis nil. Pl. Physiol.* **40**, 865–872.

Takimoto, A. & Ikeda, K. 1960 Studies on the light controlling flower initiation of *Pharbitis nil.* VI. Effect of natural twilight. *Bot. Mag., Tokyo* **73**, 175–181.

Takimoto, A. & Ikeda, K. 1961 Effect of twilight on photoperiodic induction in some short-day plants. *Pl. Cell Physiol.* **2**, 213–229.

Vince-Prue, D. 1975 *Photoperiodism in plants.* London: McGraw-Hill.

Vince-Prue, D. 1979 Effect of photoperiod and phytochrome in flowering: time measurement. In *La physiologie de la floraison* (ed. P. Champagnat & R. Jaques), pp. 91–127. Paris: C.N.R.S.

Vince-Prue, D. 1981 Daylight and photoperiodism. In *Plants and the daylight spectrum* (ed. H. Smith), pp. 223–242. London: Academic Press.

Vince-Prue, D. 1982 Phytochrome and photoperiodic physiology. In *Biological timekeeping* (ed. J. Brady), pp. 101–117. Cambridge University Press.

Vince-Prue, D. 1983 Photomorphogenesis and flowering. In *Encyclopedia of plant physiology* (N.S.) (ed. W. Shropshire Jr & H. Mohr). Berlin: Springer-Verlag. (In the press.)

Vince-Prue, D., King, R. W. & Quail, P. H. 1978 Light requirement, phytochrome and photoperiodic induction of flowering of *Pharbitis nil* Chois. II. A critical examination of spectrophotometric assays of phytochrome. *Planta* **141**, 9–14.

Wilkins, M. B. & Harris, P. J. C. 1976 Phytochrome and phase setting of endogenous rhythms. In *Light and plant development* (ed. H. Smith), pp. 399–417. London: Butterworths.

Zeiger, E., Field, C. & Mooney, H. A. 1981 Blue light responses in nature. In *Plants and the daylight spectrum* (ed. H. Smith), pp. 391–407. London: Academic Press.

Discussion

O. V. S. Heath, F.R.S. (10 *St Peter's Grove, London, U.K.*). Dr Vince-Prue's results with pulses of red light during the dark period much resemble those of Mansfield (1965) for effects of low-intensity red light in delaying the phase of a stomatal rhythm in darkness in *Xanthium pennsylvanicum*. He found that one or three interruptions of a 16 h dark period were ineffective but 64 interruptions were as effective as continuous red light. As in Dr Vince-Prue's experiment, following each period of red light with far red did not reverse the effect. For these reasons and also because maximum sensitivity was found at 703 nm, approximately midway between the main phytochrome absorption peaks, he concluded that if phytochrome was involved it was behaving very differently from its participation in the flowering behaviour of the same species (Borthwick *et al.* 1952).

References

Borthwick, H. A., Hendricks, S. B. & Parker, M. W. 1952 The reaction controlling floral initiation. *Proc. natn. Acad. Sci. U.S.A.* **38**, 929–934.

Mansfield, T. A. 1965 The low intensity light reaction of stomata: effects of red light on rhythmic stomatal behaviour in *Xanthium pennsylvanicum*. *Proc. R. Soc. Lond.* B **162**, 567–574.

Daphne Vince-Prue. This very interesting observation raises the possibility that the photo-perception characteristics that we observed for coupling with photoperiodic time measurement may be of general occurrence in time-based phenomena. It emphasizes the need to determine the wavelength sensitivity in our system.